I0493716

The First Seven Days

Limits and Continuity: Bridging the Gap between
Algebra and Calculus

General audience books by the author

Mathematical Vignettes

The First Seven Days

Limits and Continuity: Bridging the Gap between
Algebra and Calculus

Richard Poulo

Math And Such
Parker, Colorado

Copyright © 2018 by Richard Poulo

All rights reserved. No part of this publication may be reproduced, stored in a retrieval system, transmitted in any form or by any means, or translated in any form by any means without prior written permission from the author.

Library of Congress Control Number: 2018905161

ISBN: 978-0-9996174-0-3

Published by Math And Such, Parker, Colorado
www.mathandsuch.com

For my daughter Kelly and grandson Isaac

Acknowledgments

Because this book fills a pedagogical gap, numerous people reviewed it at various stages. Nick Jones, Carson Lattimore, Sergius Tunes, Steve Crandell, David Morse, Torben Hansen and Xiaolan Wang reviewed various drafts. My sincere thanks to each of them. I most especially thank my primary reviewer Tim Thomasson who made so many suggestions throughout the entire manuscript. Because of all of them this book you are reading is much improved over the book I initially wrote.

Contents

Day 1: All the Preliminaries

Each section of this book is called a "Day" because each makes a good topic for one day's lecture in a formal course or one day's reading for self-instruction. Some readers will proceed faster based on their degree of mathematical knowledge, but one Day per day is a reasonable rate of progress.

Day 1 consists of several unrelated, preliminary topics. All are here for a reason and all will be used in later Days.

Prerequisites

There are two prerequisites required to read this book.

First, one must have a reasonable knowledge of high school algebra with analytic geometry, including some comfort with working in that subject, and an understanding of what a function is. Expertise beyond that level is always useful but is not necessary.

Second, one must have a willingness to read carefully, for mathematics cannot be learned by skimming, speed reading or any other quick reading technique.

Text Notation

Text in italics emphasizes an important point in the main line of discussion.

Text in a box is outside the main line of discussion. It is generally used to review mathematical points the reader should already know and discuss isolated historical notes.

Informal Definitions

This book is about *limits*, *continuity* and *derivatives*, three concepts that will be formally and rigorously defined when the necessary groundwork has been laid. For now, informal definitions are provided so you have some idea of the concept defined by each word.

For an example of a *limit*, assume $y = x + 1$. If x is made to approach the number 1, the limit of y as x approaches 1 is the number 2. More generally, a limit of a variable is the unique number that the variable gets arbitrarily close to in response to the change in some other variable.

A function is *continuous* if there are no breaks in the (usually curved) line of the function's graph. Figure 1 shows an example of a discontinuous function.

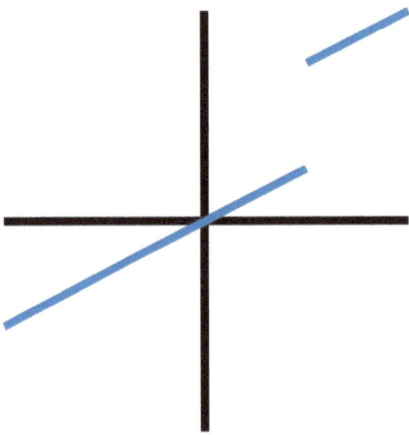

Figure 1. A discontinuous function.

The *derivative* of a function is a separate function whose value at any point is the slope of the graph of the original function at that same point. Figure 2 shows the linear function $y = \frac{1}{2}x$, which has constant slope $\frac{1}{2}$, and also shows its derivative, the constant function $y = \frac{1}{2}$. As this example shows, for linear functions this much can be done with algebra, but calculus can compute the slope function of arbitrary functions, not just linear functions.

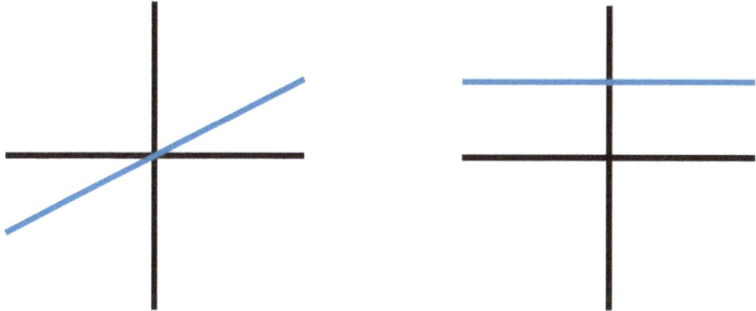

Figure 2. The function y=x/2 and its derivative y=1/2.

What is Calculus?

Calculus consists of two parts, only the first of which is discussed in this book, save in the final section A Look Ahead. This first part is called *differential calculus*.[1] Given some quantity, differential calculus deals with the rate of change of that quantity, which is called the quantity's *derivative* (it is derived from the quantity itself). Graphically the rate of change is given by the slope of the quantity's graph.

Rates of change are useful things to know. If we have the position of a car as a function of time, the derivative of the position gives us the velocity. The derivative of the velocity gives us the acceleration. The derivative of the acceleration gives us the jerkiness of the motion.

Other examples of derivatives are the rate of consumption of a chemical reactant, the rate of growth of a nation's economy and the rate of change of the number of users of social media.

---

[1] So called because its main computational equation involves taking the difference of two very closely related values.

Motivation for This Book

Just a little of the history of calculus is distributed throughout this book, always to show how difficulties in understanding arose and so emphasize why this book's exposition is important.

Calculus was created about the year 1665. Its two independent creators, Isaac Newton and Gottfried Leibniz, never in their lives defined its basic concepts such as the derivative quite right. They each proceeded because their intuition and their results told them they were on the right track.

Succeeding generations of mathematicians did no better, one great mathematician (Jean d'Alembert) telling his students to have faith, that certainty about calculus would come with practice. Other mathematicians lacked such faith, felt calculus was invalid and that it produced correct results only because errors were somehow cancelling each other. In the earliest formulations this was actually the case, which illustrates just how difficult it was to rigorously formulate calculus.

This lack of a rigorous formulation of calculus continued for *two centuries* after its creation. Not until the 1820s did the French mathematician Augustin-Louis Cauchy rigorously formulate what are now called *limits* (rigorously defined in Day 5) and *continuity* (rigorously defined in Day 6), use them to create an almost completely rigorous basis for calculus, and earn mathematical immortality for his work.[2] It was more decades before the German mathematician Karl Weierstrass finally completed the process and created the fully rigorous, modern formulation presented in this book.

---

[2] While too advanced for newcomers to calculus, an excellent reference is the book The Origins of Cauchy's Rigorous Calculus by Judith V. Grabiner. This meticulously researched work describes the situation in mathematics before Cauchy and how Cauchy brought about a new normal in the foundations of calculus. It was republished by Dover in 2005.

Given the span of time it took mathematicians to get it right, students should feel no pang of stupidity for having their own trouble learning about limits and continuity. This is good news, for you must learn these two prerequisite concepts. If you do not learn them well, the penalty is that concepts in calculus itself, all of which build on these prerequisites, may become impossibly difficult due to confusions and omitted explanations.

Limits and continuity are not difficult given a step-by-step exposition of them in all aspects. Unfortunately, in an eagerness to get to calculus itself, they are often covered briefly and in rapid succession with too little explanation, too little perspective and too few examples. To cite one instance, the important Equation of Continuity (defined in Day 6) sometimes gets little more than a passing mention. The result of insufficiently covering limits and continuity is that almost everyone has trouble learning calculus and so "knows" calculus is hard.

This book dwells on and clarifies limits and continuity, then concludes by defining derivatives. After learning from it (it is written to allow self-study as well as being used as a classroom text) the student can continue with any standard calculus text to master this subject.

Mathematical Self Study

Many years ago, as an undergraduate, I was exposed to the first italicized paragraph below, which prompted my initial skeptical reaction, "Yeah, right!" Quite some time later, when bogged down in a textbook, I recalled this statement and tried it because I had nothing else left to try. To my amazement it worked although in retrospect it became obvious that it would. Since then I have used it whenever appropriate and it always came through.

In reading this book, be very, very certain you never go past a word or symbol you do not fully understand. If the material becomes confusing or you can't seem to grasp it, there will be a word or symbol just earlier that you have not understood. Don't go any further, but go back to BEFORE you got into trouble, find the misunderstood word or symbol and get it defined.

To a student new to this idea, the task of finding a misunderstood word or symbol might be more difficult than anticipated. Here is an easy way to find it. Go back to before you got into trouble, then read forward slowly, word by word. You will find you understand perfectly until you reach a first word where something does not seem quite right or you do not quite understand the concept. That first word is the word you seek.

It is a misunderstood word even if you know the word in other contexts, for it may have a slightly different meaning in the current context. You will know you completely understand the word when you can resume reading without any sense of less than complete understanding.

This process gets easier with practice.

All this is especially important in the study of mathematics in which each topic carefully builds on the previous topic. To help with this, a number of words and symbols have been defined in the text. There is also a table of the symbols used in this book (page 86) and a glossary (page 87). The glossary is not limited to mathematical words but includes a few commonly misunderstood English words and abbreviations such as "i.e." and "e.g."

A Descriptive Introduction to Calculus

Calculus assumes that limits and continuity have already been defined and starts by defining the derivative.

Look at the function $y = f(x)$ graphed as the black curve in Figure 3. The derivative of $f(x)$ is another function in its own right. Its value at x is the slope of the curve of $f(x)$ at x, which tells us the rate at which y changes with respect to x at that particular point. In other words, the derivative tells us, at any given point x, whether y is changing 3 times as fast as x is changing, 5 times as fast or perhaps only half as fast as x is changing.

Because it is hard to define the slope of a curve at a given point in a direct way, we define the slope of a curve to be the slope of the tangent line at that point. The blue lines in Figure 3 show tangents at different points.

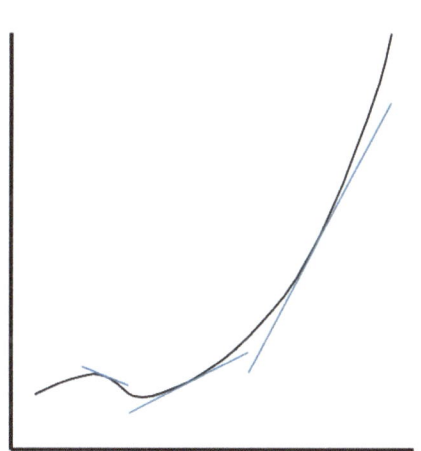

Figure 3 Graph of a function and some of its tangent lines. Each tangent line touches the curve at only one point.

Even finding the slope of a tangent is quite a difficult task for which to devise a means of computation. In algebra one is given a linear equation and then one graphs the line. However, now one is given the line as the tangent to a curve and one must determine the linear equation, which then gives us the slope. This is a definite break from algebra, which does not offer mathematical tools capable of determining the equation from the tangent line.

Limits and continuity are the new mathematical tools required to achieve this. When both limits and continuity are thoroughly clear, we will put them together to compute the slope of the tangent line, which is the derivative.

Actually, Limits Are Optional

The discussion so far makes it sound as though limits are necessary to the study of calculus. In the standard formulation of calculus limits are indeed necessary, yet there is a so-called *non-standard* formulation that defines derivatives without using limits.

Nothing comes free. Instead of defining limits the non-standard formulation defines *infinitesimal numbers*. These are numbers with a seemingly impossible description, for the positive infinitesimals (for example) are greater than zero but less than any positive number no matter how small. Similarly for the negative infinitesimals.

Figure 4 shows the number 0 with its zero-length sets of the infinite number of positive and negative infinitesimals that fit into the number line on each side of 0. Because infinitesimals can be added to any ordinary number, not just 0 but every ordinary number is surrounded by its own new sets of numbers involving infinitesimals.

The apparent contradiction that makes infinitesimals sound impossible can, in fact, be resolved. Infinitesimal numbers can be rigorously defined using advanced set theory, *which we will not be doing*. Once infinitesimals have been defined, a derivative (remember that the value of a derivative is the slope) is simply defined as a ratio of two infinitesimals, just as the slope of a straight line is defined as the ratio of two ordinary numbers.

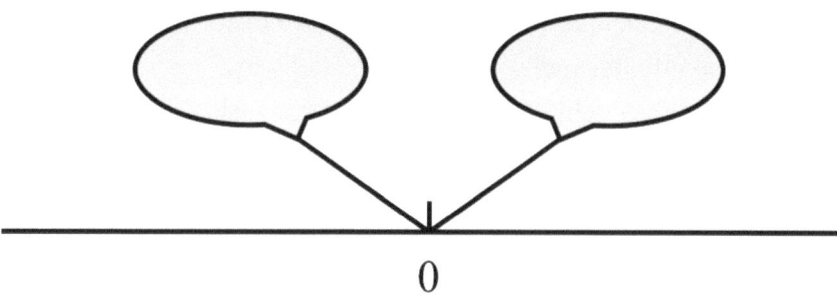

0

Figure 4. The number 0 surrounded by the negative and positive
infinitesimals, shown expanded via the balloons.

Leibniz's non-rigorous intuition about infinitesimals
motivated his notation $\frac{dy}{dx}$ (read "d y over d x") for the derivative, for
he considered dy and dx to be infinitesimal numbers whose ratio was
taken. Today derivatives are standardly defined using limits and yet
the notation $\frac{dy}{dx}$, which misleadingly suggests a ratio of
infinitesimals, is still used. This will be discussed in Day 7.

It was not known that there could be two formulations until
1961, *three* centuries after the invention of calculus, when the
rigorous formulation based on infinitesimals was created by the
German-American mathematician Abraham Robinson.[3] As a result,
over these three centuries the concepts of the two formulations were
mingled causing confusion regarding not only notation as cited above,
but causing confusion regarding the ideas. One of these confusions

---

[3] One good introduction to infinitesimals and the non-standard formulation of calculus is the
book Infinitesimal Calculus by Henle and Kleinberg. Containing both mathematical and
historical discussions, it is within the reach of students who have completed a first year of
calculus. It was republished by Dover in 2003.

you may have already encountered is that calculus is frequently called "infinitesimal calculus" even when it is developed using limits and there are no infinitesimals.

This book will plainly present the standard formulation of calculus based on limits while pointing out where confusions between the formulations are to be avoided.

It must be pointed out that regardless of whether limits or infinitesimals are chosen for the foundation, the superstructure of calculus built on the foundation is exactly the same.

Quick Review of Functions and Formulas

Students often confuse *function* with *formula*.

A *function* $f(x)$ is just a correspondence (the usual mathematical term is "mapping") from one set of numbers to another set of numbers that satisfies a special property to be specified in a moment. An example of a correspondence that is a function is $y = x + 1$, alternatively written $f(x) = x + 1$. (We freely switch back and forth between the two notations $y = expression$ and $f(x) = expression$ as convenient.) Given the number 2 as the value of x the value of this function is 3. Given the number 17 the value of the function is 18. Simple enough.

A function must have the special property that for each value of x there corresponds at most one value of y. An example of a correspondence that is not a function is $y = \pm\sqrt{x}$, which for some values of x (those greater than 0) has two values of y. In particular, if $x = 4$ then $y = +2$ and $y = -2$.

It is allowed for different values of x to correspond to the same value of y. It is only required that each value of x itself corresponds to (at most) one value of y.

A *formula* is an expression that has a numeric value when all its variables are assigned numbers. The function $y = x + 1$ is defined by the formula $x + 1$.

But some functions are not defined by formulas in this way as in this next example. We define a somewhat contrived function below. The two conditions jointly define the single function $f(x)$.

$f(x) = x$ if x is zero or a positive number

$f(x) = x^2$ if x is a negative number

The function is graphed in Figure 5.

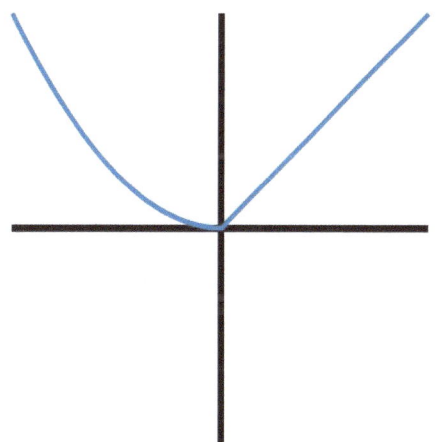

Figure 5. A function that is not defined by a single formula.

The definition of this function does not provide a single formula whose value is x when x is zero or positive and whose value is x^2 otherwise. Perhaps you can write a single formula for this function, but this is still a function regardless of whether we have such a formula.

In this book we will use simple functions for which there is no single formula. The correspondences (or mappings) defining the functions may be described by multiple formulas but they are still perfectly valid functions.

All problems in this book are simple. Students should actually do them. All answers are provided in a section at the end (page 72).

Problems

1. Is this a function? Note that the two formulas together define a single function.

    $$y = x \qquad \text{if } x \geq 0$$

    $$y = -x \qquad \text{if } x < 0$$

2. Is this a function?

    $$y = x \qquad \text{if } x \geq 0$$

    $$y = -x \qquad \text{if } x \leq 1$$

3. Is this a function?

    $$y = x \qquad \text{if } x \geq 0$$

    $$y = -x \qquad \text{if } x \leq 0 \qquad \text{note } \leq \text{ whereas problem 1 has } <$$

Day 2: What Is the Problem?

Some Simple Functions

Look at Figure 6 which graphs the function

$$y = x + 1$$

Even ignoring the fact that the graph is a straight line, this is still a "nice" function. It has no corner, no roughness and no breaks. There isn't much to say about this function – yet.

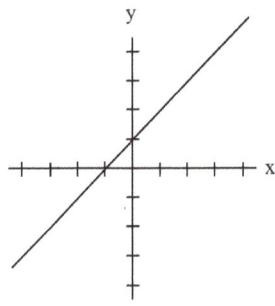

Figure 6 The function $y = x + 1$

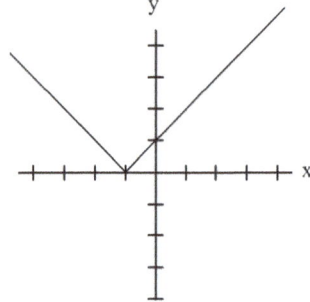

Figure 7 The function $y = |x + 1|$

Now look at Figure 7 which graphs the absolute value of the function in Figure 6, i.e., it graphs the function $y = |x + 1|$ which is given by

$$y = x + 1 \qquad \text{if } x + 1 \geq 0$$

$$y = -(x + 1) \qquad \text{if } x + 1 < 0$$

14

This figure contains a bent line showing the absolute value of $x+1$. This function is less nice in that it has a sharp corner. For our current purpose this function is still too nice.

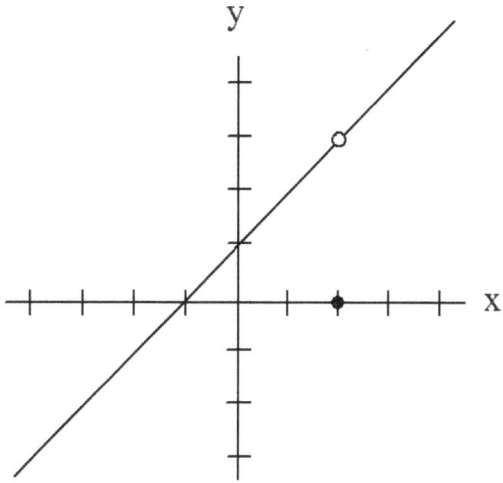

Figure 8. $y=x+1$ except that $y=0$ when $x=2$

The graphical notation of a small, filled in circle ⊶ represents a single geometrical point. Such a point has zero size but of necessity we represent it as a filled in circle.

The graphical notation of a hollow circle on a line ⟋ represents a single geometrical point missing from the line.

Finally look at Figure 8 which graphs the function

$$y = x+1 \qquad \text{if } x \neq 2$$

$$y = 0 \qquad \text{if } x = 2$$

This function is unlike anything in algebra but functions such as this occur repeatedly in calculus. In order to find slopes on a curve, one must be able to solve the problem shown in this figure.

This graph is a line except that one point has been pulled down below the line onto the x-axis. It represents a function that is even less nice than the previous functions because of the break in its graph. Because of the break this function is not "continuous", a word for which we will eventually give a precise mathematical meaning. For now, we will use our intuition since it is clear from its graph that the function is not continuous.

In order to be continuous this function should have the value 3 when $x = 2$, but how do we know that? Of course the formula $x+1$ takes the value 3 when $x = 2$, but we are not supposed to use the formula $x+1$ when $x = 2$. In addition, despite the obvious conclusion from the graph that 3 is the value that would make the function continuous, visual representations are not proofs. So let us pretend that we do not know how to compute $x+1$ when $x = 2$ because when we have formulas more advanced than $x+1$ this will actually be the case.

Looking at the two formulas defining the function and remembering that $x+1$ is not to be used when $x = 2$, is there still a way to extract 3 as what would have been the "nice" value of the function, the value that would have made it continuous when $x = 2$?

This turns out to be the key question. Getting to its answer will drive us down the path to calculus.

Figure 8 and the function graphed in it will be referred to throughout this book.

16

Problems

1. Recall that a linear function, as defined in basic algebra, is any function of the form $y = mx + b$ where the constants m and b may be any numbers at all. For example, $y = 2x + 3$ and $y = -17x - 1000$ are linear functions. Assuming $m \neq 0$ how many roots does a linear function have? (A *root* is any value of x that makes $y = 0$.)

2. If m is not restricted to be nonzero how many roots might a linear function have?

3. How many roots does the function in Figure 8 have?

   What are the root(s)?

   Is this function linear?

A Shot in the Dark

Looking again at Figure 8, a reasonable guess is that the nice value is 3 because that is the only value that y never assumes. However, Figure 9 and Figure 10 show two other functions that should have the value 3 at $x = 2$. The function in Figure 9 not only never assumes the value 3 but also never assumes various other values. In particular, y in Figure 9 never assumes any of the values 1, –2, ½, 5 or any of many other values.

From this it should be clear we cannot pick the number 3 as the nice value for Figure 8 simply because it is the only value y never assumes. Note that the graph in Figure 9 was designed to match the graph in Figure 8 when x is in the interval [1, 3].

The notation $[a,b]$ defines an interval, which is a set of numbers. In particular, it is the set of numbers x such that $a \leq x \leq b$. Note that the endpoints of the interval are included in the interval. Compare this notation with the interval notation (a,b) which excludes the endpoints and means all numbers x such that $a < x < b$. Mixed notation may be used, e.g., $[a,b)$ for $a \leq x < b$.

Both $[a,b]$ and (a,b) may be read as "the interval from a to b".

When one wishes to convey information about the endpoints, one may read $[a,b]$ as "the closed interval from a to b" and read (a,b) as "the open interval from a to b".

On occasion in this book, the notation (a,b) does not have its usual meaning of an open interval on the number line but instead means the x-y coordinates of a point in the plane. Context should make clear which meaning is intended.

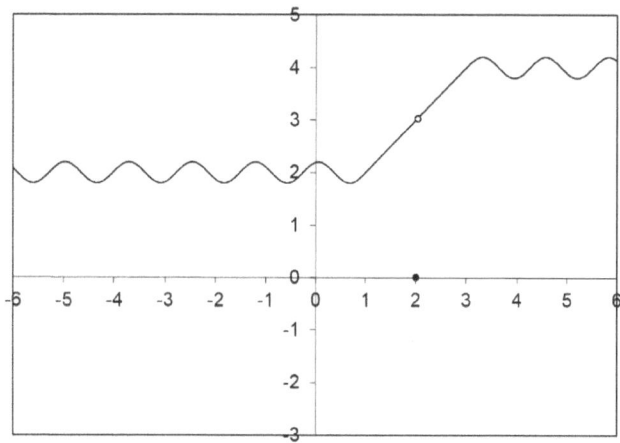

Figure 9 This function matches the function of Figure 8 in the interval [1,3].

Figure 10 shows another graph designed to match Figure 8 for *x* in the interval [1,3], but in this graph *y* assumes all values including the value 3. In fact, *y* = 3 for two values of *x*, namely *x* = 0 and *x* = 4.

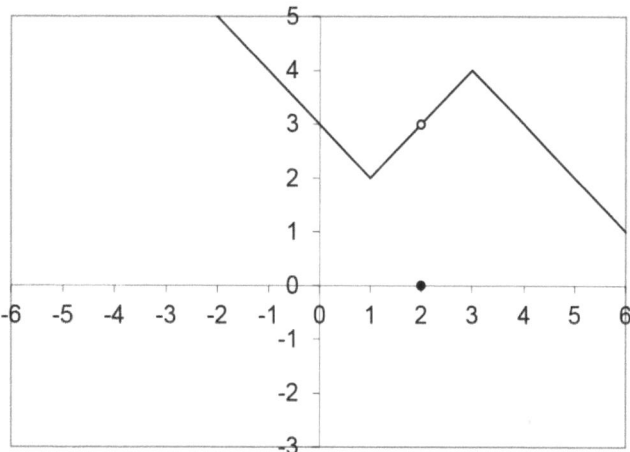

Figure 10 Another function matching the function of Figure 8 in the interval [1,3].

Our conclusion is that what happens outside the interval $[1, 3]$ cannot determine what happens inside that interval. Rephrasing this conclusion gives the following.

What happens far away from $x = 2$ cannot determine the desired value at $x = 2$.

The symbol $\in$ when used in a statement such as $x \in [1,3]$ means x assumes a value in the interval $[1,3]$, i.e., $1 \le x \le 3$. Similarly, the statement $x \in (1,3)$ means x assumes a value in the interval $(1,3)$, i.e., $1 < x < 3$.

"$x \in [1,3]$" can be read "x is in the interval 1 to 3".

Problems

1. Draw at least one function that matches Figure 8 in the interval [1, 3] but differs from it outside that interval. See how inventive you can be. Remembering that not all mappings are functions be sure that you have drawn functions.

2. The single mapping defined by these three formulas taken together is intended to be a function. Is this actually a function?

$$y = x+1 \qquad \text{if } x \in [1, 3] \text{ but } x \ne 2$$

$$y = 0 \qquad \text{if } x = 2$$

$$y = 3 \qquad \text{if } x = 4$$

Graph it. Note that y assumes all values in the interval [2, 4], this interval being along the y axis not the x axis.

How Close Must We Get?

If what happens far away from $x = 2$ does not matter, just how far away is "far away"? Does $x = 1$ qualify as far away? If $x = 1$ is close, the function value at $x = 1$ would determine (or help determine) the nice function value at $x = 2$.

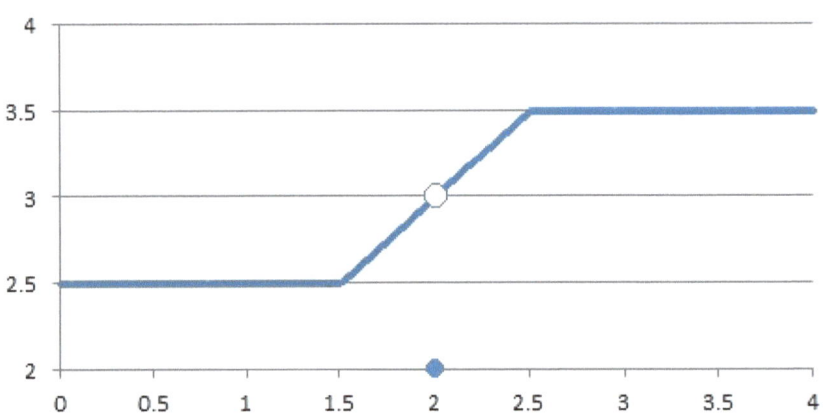

Figure 11 A function matching the function of Figure 8 in the interval [1.5, 2.5].

Figure 11 shows the graph of the function

$$y = 2.5 \qquad \text{if } x < 1.5$$

$$y = x + 1 \qquad \text{if } x \in [1.5, 2.5] \text{ but } x \neq 2$$

$$y = 0 \qquad \text{if } x = 2$$

$$y = 3.5 \qquad \text{if } x > 2.5$$

At $x = 1$ we have $y = 2.5$ yet this function still equals the function in Figure 8 for $x \in [1.5, 2.5]$ and in particular for $x = 2$. So $x = 1$ is too far away to determine the nice value at $x = 2$.

Let's try getting closer. Does $x = 1.9$ qualify as far away? Figure 12 and Figure 13 show that this is far away also, for even the following function with multiple discontinuities still looks like the original function between 1.95 and 2.05.

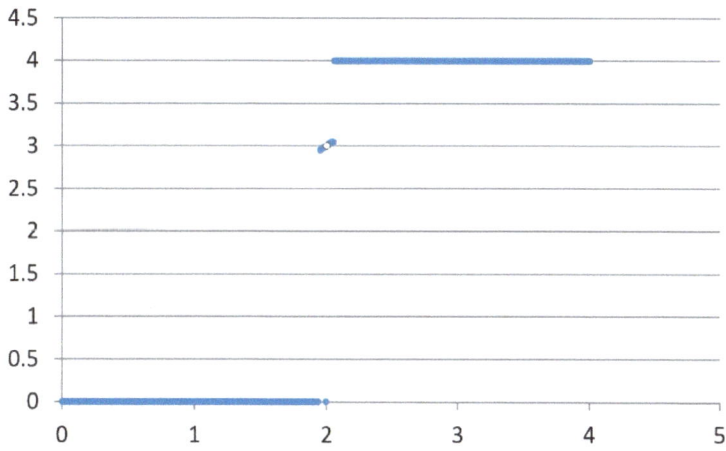

Figure 12 A multiply discontinuous function matching the function of Figure 8 in the interval [1.95, 2.05]. The middle section of the graph is a very short line segment with a single point missing. That missing point appears as an isolated point on the *x* axis.

Figure 12 shows the graph of the function

$y = 0$ if $x < 1.95$

$y = x + 1$ if $x \in [1.95, 2.05]$ but $x \neq 2$

$y = 0$ if $x = 2$

$y = 4$ if $x > 2.05$

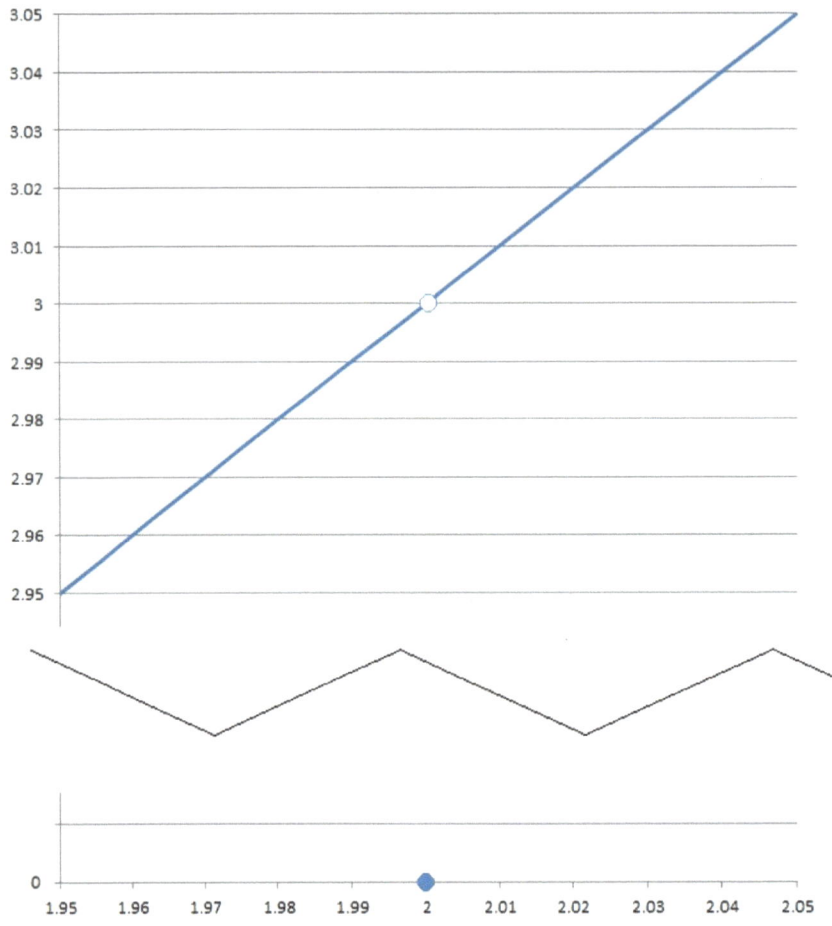

Figure 13 A blowup of Figure 12 showing the interval [1.95, 2.05].

We have moved closer to $x = 2$ yet we have not been able to affect in the slightest way the function value that is actually at $x = 2$. Not even other function discontinuities as shown in Figure 12 make any difference.

24

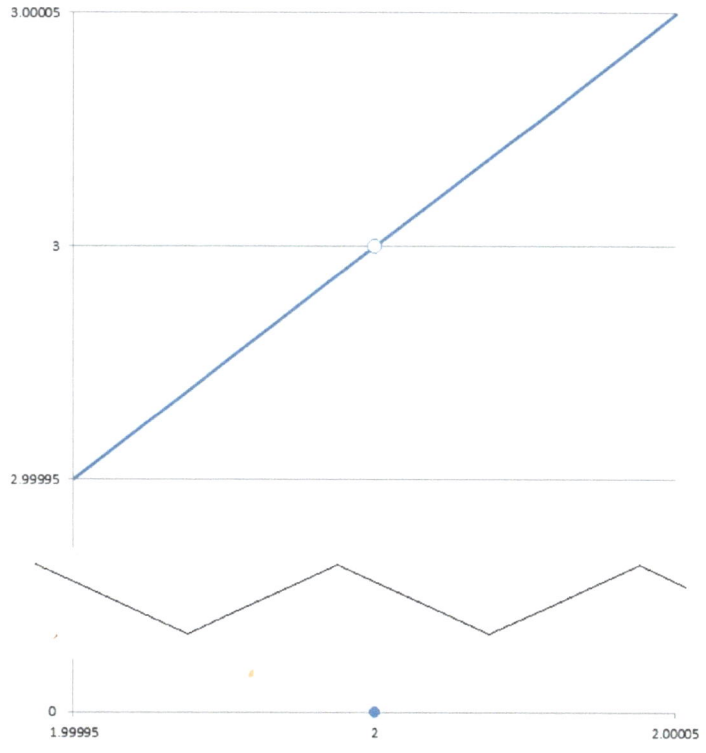

Figure 14 A function matching the function of Figure 8 in the interval [1.99995, 2.00005], showing only that interval.

Figure 14 shows the graph of the function

$y = 0$ if $x < 1.99995$

$y = x + 1$ if $x \in [1.99995, 2.00005]$ but $x \neq 2$

$y = 0$ if $x = 2$

$y = 4$ if $x > 2.00005$

Let's get closer still. How about $x = 1.9999$? The blowup in Figure 14 shows that even this is far away, for even this function looks like the original function between 1.99995 and 2.00005.

In fact, no matter how close we get to $x = 2$, we are still far away. We arrive at the following conclusions.

The value of y at any point other than $x = 2$ does not determine the nice value of y at $x = 2$ (the value that would make the function continuous).

Nor can we use the value of y at the point $x = 2$ to determine the nice value of y because that gives the value 0, not 3.

Day 3: Developing a New Intuition

The First Part of the Answer

Still referring back to the function graphed in Figure 8 on page 15, it is time to start developing a solution to the problem of finding the value of the function at $x = 2$ that makes the function continuous at that point. The solution is not difficult but from Day 2 we now expect that it will be slippery. (Don't forget that it stumped Newton and Leibniz).

Here is the idea stated without mathematical rigor (rigor is for Days 4 and 5):

Since we can't use the function value at any one point, we must see what happens to the whole set of function values as we approach $x = 2$ without actually reaching $x = 2$.

It is ok if that is a bit confusing, for the mission of this section is to make it clear by using examples.

For this first half of the answer, we want to see if the function has this property:

y approaches 3 as x approaches 2.

Let's check our intuition. Does y really approach 3 no matter how close x gets to 2?

As x approaches 2, written $x{\rightarrow}2$ and read "x approaches 2", y is always evaluated by $y = x + 1$ because x is never equal to 2. (Remember that when x does equal 2, y is evaluated by $y = 0$.) As $x{\rightarrow}2$, y always stays 1 greater than x and so $y{\rightarrow}3$. It seems that our intuition is correct, that y really does approach 3.

The Second Part of the Answer

Now for the second check on our intuition, one we will spend a little more time on. Is 3 the only value that y approaches? Doesn't y approach 2.9 also? Let's make $x \rightarrow 2$ from below, i.e., x starts out smaller than 2, say 1.5, and gets bigger, passing through all numbers between 1.5 and 2, including for example the numbers 1.7, 1.8 and 1.9 on its way to 2.

As $x \rightarrow 2$ from below we find that x first approaches 1.9, so as $x \rightarrow 1.9$ it seems that $y \rightarrow 2.9$, i.e., y keeps getting closer to 2.9. But when x equals and then passes 1.9 on its way to 2, we find that y equals and then passes 2.9 on its way to 3, ultimately moving away from 2.9, not toward 2.9. So we cannot say that $y \rightarrow 2.9$ as $x \rightarrow 2$. (We can say that $y \rightarrow 2.9$ as $x \rightarrow 1.9$, but not as $x \rightarrow 2$.)

Perhaps 2.9 is a number too far away from 3. Let's try 2.999. Does $y \rightarrow 2.999$ as $x \rightarrow 2$ from below? We find the same thing happening. y approaches 2.999 at first. However, as x approaches, passes through and then gets farther away from 1.999 on the small remainder of its journey towards 2, we find that y approaches, passes through and then gets farther away from 2.999 on the small remainder of its journey towards 3.

This is encouraging but in this subject we must be careful. Let us ask again. Is 3 really the only number that y approaches as $x \rightarrow 2$? We have not yet considered numbers greater than 3. As $x \rightarrow 2$ from below, y is not only approaching 3 but is also approaching 4, 5, 7 and 1000. y never gets as close to 4, 5, 7 or 1000 as it does to 3 but it is still approaching each of these other numbers. A similar statement holds when $x \rightarrow 2$ from above in which case y is not only approaching 3 but is also approaching 2, 0 and -17.

We must capture the idea that the distance between y and 3 can be made as small as we like simply by making x sufficiently close to 2. If we can do this, we can say that 3 *is the only number y approaches as x→2. By "approaches" we now mean "approaches arbitrarily close to".*

A Detour for Notation

As you studied mathematics you learned new notations. For example, you learned the cube root notation $\sqrt[3]{x}$. This notation indicates a numeric value, namely the number whose cube is x. If $x = 27$, the value of the notation $\sqrt[3]{27}$ is the number 3 and it may be used as such in equations. For example,

$$4 = \sqrt[3]{27} + 1$$

In other words, the notation $\sqrt[3]{x}$ is just a handy way of writing a number when we do not or cannot write the number directly. As another example, $\sqrt[3]{13}$ is a certain number which happens to be an infinite decimal. The notation lets us write this number exactly without writing out an infinite number of digits.

Now you must learn a new notation that also specifies a number and is also used when we are unable to write the number directly:

$$\lim_{x \to 2} y$$

"lim" stands for the word "limit". Just like the cube root notation, this limit notation has a numeric value that we can use in equations.

Now you have seen the new notation but its meaning has not been defined. We must define it before using it but readers have probably guessed the meaning by now. Based on intuition only, we have found the method of getting 3 as the nice value of the function when $x = 2$. The method deals with the limiting values of x and y, which are the values these variables approach.

As a notation we could write

$$\text{as } x \rightarrow 2, \ y \rightarrow 3$$

which is read "as x approaches 2, y approaches 3". But we always want to emphasize the limiting value of y because that is more important than the limiting value of x. After all, we *choose* the limiting value of x. It is just the number we make x approach, for example, it is the number 2 when $x \rightarrow 2$. However, we want to *discover* the limiting value of y. That is the whole purpose of the limit concept.

For this reason, we rewrite the above as an equality:

$$\text{as } x \rightarrow 2 \text{ the limiting value of } y = 3.$$

Our final notation for this is

$$\lim_{x \rightarrow 2} y = 3$$

This is read "the limit of y (as x approaches 2) equals 3." You can include or exclude the parenthesized words as you prefer in any given context. This means no more than our previous notations but turns out to be far more convenient because it is written as an equality and can be used in equations.

Despite the above elaboration we still have not provided a rigorous definition of a limit so the notation does not yet have a rigorous meaning. When we do rigorously define limits in Day 5 and when the numerical value of this new notation

$$\lim_{x \rightarrow 2} y$$

is actually computed in Day 7, the result really will be 3.

Day 4: Limits: A First Attempt at Rigor

This is the most difficult Day. Read carefully, for comprehension is required in order to finish this book. Review the advice in the section Mathematical Self Study on pages 5-6.

Overview

Now let's try making our intuition about limits rigorous. We want to capture the idea that as x gets close to 2, y gets close to 3. Part of the answer lies in measuring how close x and y are to the respective numbers they are approaching.

For no compelling reason, the Greek letters δ (delta) and ε (epsilon) are universally used for this purpose. The fact that they are unfamiliar should not put one off, they are just fancy ways of writing the letters d and e. δ is used for the distance of x from the number 2 while ε is the distance of y from the number 3.

$$\delta = |x - 2|$$

$$\varepsilon = |y - 3|$$

Using the first of these equations as an example, it is read "delta equals the absolute value of x minus 2."

We choose to use absolute values because it turns out we will want δ and ε to be positive numbers regardless of what side of 2 the variable x happens to be on, i.e., regardless of whether x approaches 2 from below (1.9, 1.99, 1.999, …) or from above (2.1, 2.01, 2.001, …).

Our intuition guides our formulation of a rigorous definition of the following mathematical statement.

$$\lim_{x \to b} y = c$$

Intuition says that as $x \to b$ it should force $y \to c$. Figure 15 and Figure 16 show our intention graphically. Now we have to find a mathematically precise definition that captures that intention.

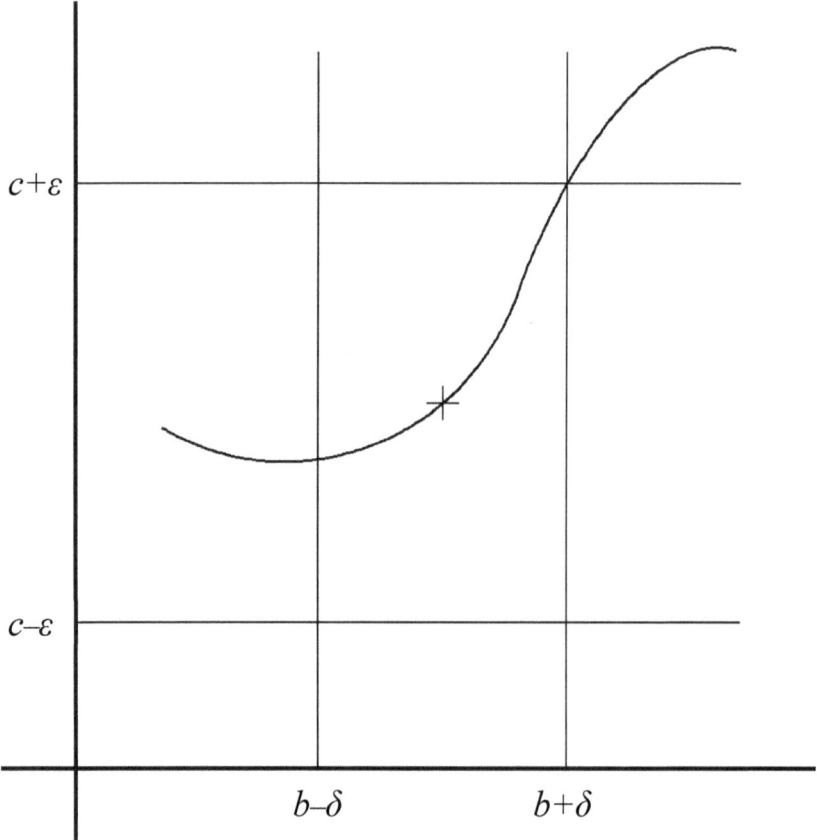

Figure 15 When x is within δ of b, y is guaranteed to be within ε of c.

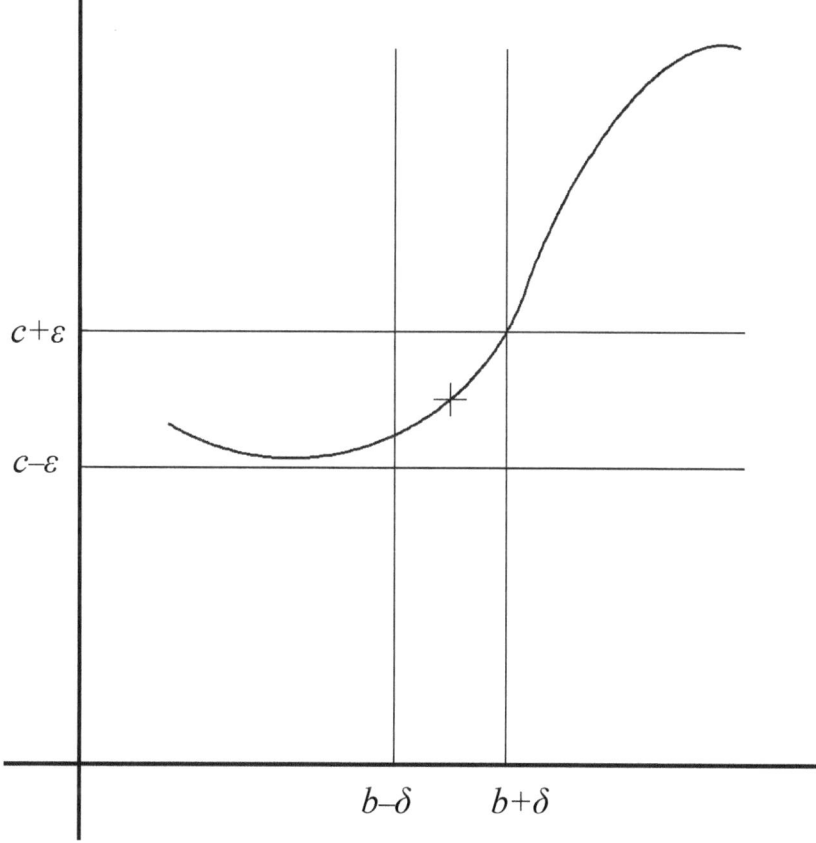

Figure 16 The same as the previous figure but with smaller δ and ε. The intention is that both get smaller so that as x closes in on b, y closes in on c.

Our first definition is the most obvious one of its kind but *it will turn out not to do what we want*. We try it anyway to see what goes wrong so we can learn how to fix it.

Trial definition (does not work). A function $f(x)$ is said to have a *limit* c as $x \to b$, written $\lim\limits_{x \to b} f(x) = c$, if given any number $\delta > 0$ we can find some $\varepsilon > 0$ (which will depend on δ) such that

whenever $\qquad |x - b| < \delta$

then $\qquad |y - c| < \varepsilon$

The intention is that δ and ε become smaller and smaller, getting arbitrarily close to 0. As long as x is within δ of b, y must be within ε of c. It does not matter if, as x approaches b, y briefly moves away from c, as long as y stays within the current ε of c.

We will give this trial definition a trial run using a simple function, the linear function $f(x) = 2x - \frac{1}{2}$, shown in Figure 17. We know $f(1) = \frac{3}{2}$. Because the graph of $f(x)$ is a straight line with no breaks, our intuition tells us that the limiting value should be on that line. This is why we have chosen a linear function to start with. Everything about them is simple. So we expect to find that

$$\lim\limits_{x \to 1} f(x) = \tfrac{3}{2}$$

If this is not true, the definition does not capture our intention. We now try to validate our trial definition by showing that

- $\frac{3}{2}$ satisfies the definition of $\lim\limits_{x \to 1} f(x)$ and

- no other number also satisfies the definition.

Does the Expected Limit Satisfy the Definition of a Limit?

First we check that $\frac{3}{2}$ satisfies the definition of the limit. This means $c = \frac{3}{2}$ in the definition. Because $x \to 1$, we also have $b = 1$ in the definition. The two inequalities become

$$\text{whenever} \qquad |x-1| < \delta$$

$$\text{then} \qquad \left|y - \tfrac{3}{2}\right| < \varepsilon$$

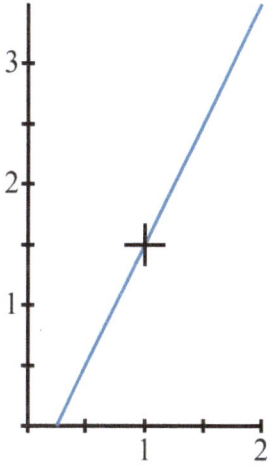

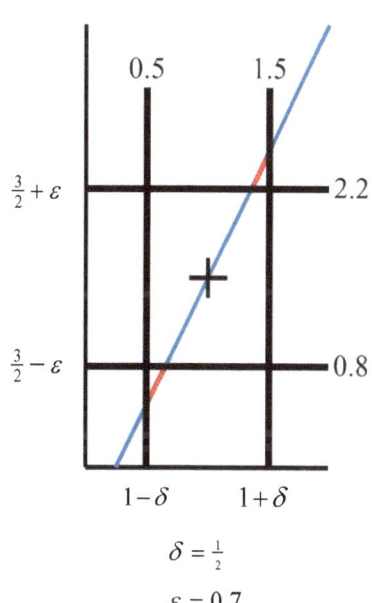

$$\delta = \tfrac{1}{2}$$

$$\varepsilon = 0.7$$

Figure 17 The function $f(x) = 2x - \frac{1}{2}$ showing the point $\left(1, \frac{3}{2}\right)$, which is marked by the cross.

Figure 18. Initial δ-ε diagram to see if $\lim\limits_{x \to 1} f(x) = \frac{3}{2}$ according to the trial definition. The red segments show that this value of ε (0.7) is too small.

We start by selecting $\delta = \frac{1}{2}$. Can we find an ε so that the definition of a limit is satisfied? First we try $\varepsilon = 0.7$ as shown in Figure 18. This fails, for when x is near either end of the interval $(1-\delta, 1+\delta)$ the value of y is on one of the line segments drawn in red, i.e., y is outside the desired y interval $(\frac{3}{2}-\varepsilon, \frac{3}{2}+\varepsilon)$.

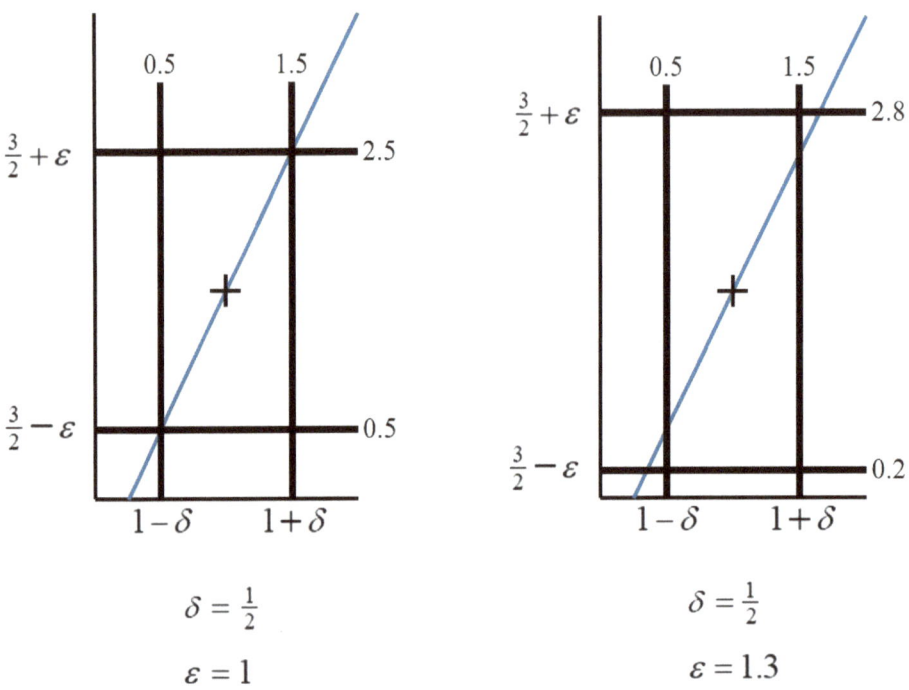

$$\delta = \frac{1}{2} \qquad\qquad \delta = \frac{1}{2}$$

$$\varepsilon = 1 \qquad\qquad \varepsilon = 1.3$$

Figure 19 a & b. δ-ε diagrams to see if $\lim\limits_{x \to 1} f(x) = \frac{3}{2}$ according to the trial definition. By comparing to Figure 18, one can see in Figure a that $\varepsilon = 1$ is the smallest acceptable value of ε for $\delta = \frac{1}{2}$. The second value of ε (1.3) in Figure b was arbitrarily chosen as representative of values greater than 1.

Let us try $\varepsilon = 1$ as in Figure 19a, where by comparison with Figure 18 it is visually clear that 1 is the minimum acceptable value of ε for $\delta = \frac{1}{2}$. But while visual clarity is nice, it is not a proof. Instead, we must demonstrate that for $x \in \left(\frac{1}{2}, \frac{3}{2}\right)$, i.e., $|x-1| < \frac{1}{2}$, it is the case that $\left|y - \frac{3}{2}\right| < 1$.

The statement $x \in (a,b)$ can always be expressed as an inequality. Let m be the midpoint of the interval (a,b) and let d be half the length of the interval, i.e., $m = \frac{1}{2}(a+b)$ and $d = \frac{1}{2}(b-a)$. Then the statement $x \in (a,b)$ means that x is no farther from the midpoint m than the distance d, which in symbols is $|x - m| < d$.

In the other direction, if we are given $|x - m| < d$, we can write $a = m - d$ and $b = m + d$, then immediately conclude that $x \in (a,b)$.

Switching between these two equivalent statements will be referred to as "restating in different notation".

Using the function that we have chosen, $y = 2x - \frac{1}{2}$, this last inequality is equivalent to

$$\left|2x - \tfrac{1}{2} - \tfrac{3}{2}\right| < 1$$

Simplifying yields

$$\left|2x - 2\right| < 1$$

Dividing by 2 finally gives us

$$\left|x - 1\right| < \tfrac{1}{2}$$

which is what we are assuming is true. Using this assumption as a starting point and working backwards we derive $\left|y - \frac{3}{2}\right| < 1$, which is what we desire.

Given the conclusion that $\left|y - \frac{3}{2}\right| < 1$ it is clearly also true that $\left|y - \frac{3}{2}\right| < 1.3$, where now $\varepsilon = 1.3$ as shown in Figure 19b. In fact, any $\varepsilon \geq 1$ works for $\delta = \frac{1}{2}$.

So for this particular value of δ we have found a suitable ε (in fact, an infinite number of them) and so far have satisfied the trial definition. But the definition requires that, given any δ whatsoever, however small, we must be able to find a suitable ε.

Let us now take $\delta = 0.01$. A strange thing happens, namely that the same ε still works, for if

$$|x-1| < 0.01$$

then it is certainly true that

$$|x-1| < \tfrac{1}{2}.$$

When this is true we already know that

$$|y - \tfrac{3}{2}| < 1$$

So the same value of ε, specifically $\varepsilon = 1$, works for $\delta = 0.01$ and, by the same reasoning, works for all smaller δ.

This is unexpectedly easy, too easy in fact, and will later lead to trouble. For now, we have satisfied the trial definition and have shown that $\tfrac{3}{2}$ meets the definition's δ-ε criterion for being the value of $\lim_{x \to 1} f(x)$. So far the trial definition is doing what we want.

Does Any Other Number Also Satisfy the Definition?

It remains to show that no other number also satisfies our trial definition. Instead of $\tfrac{3}{2}$ we now attempt to show that 1 satisfies the definition of $\lim_{x \to 1} f(x)$. Of course, we want this attempt to fail.

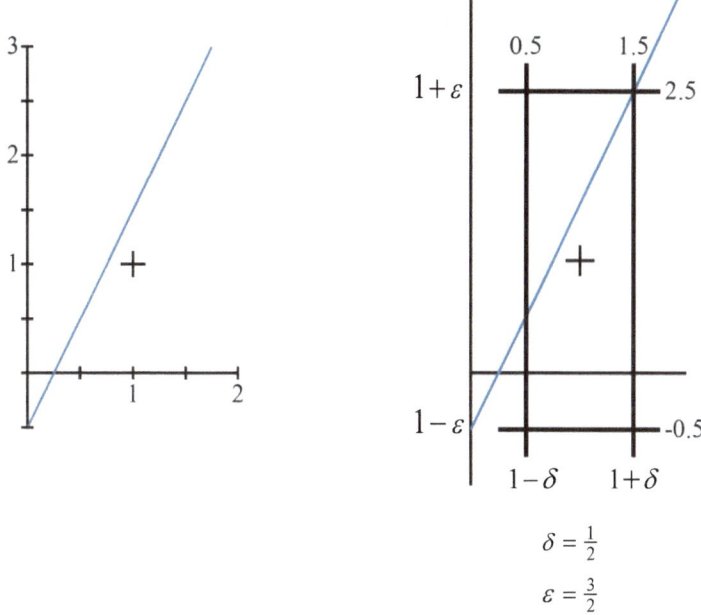

$$\delta = \tfrac{1}{2}$$

$$\varepsilon = \tfrac{3}{2}$$

Figure 20 a) The same function $f(x) = 2x - \tfrac{1}{2}$ is being examined. This time the cross marks the point (1,1).

b) δ-ε diagram to see if $\lim_{x \to 1} f(x) = 1$ according to the trial definition. This diagram shows why the number 1 (as a value of y) should not satisfy the definition of a limit as $x \to 1$, for the point (1,1) is separated from the line.

Figure 20 shows visually why we want the number 1 to fail to satisfy the definition of a limit. But the math will show that it does satisfy our trial definition. We will see that not only does 1 satisfy the δ-ε criterion for a certain δ ($\delta = \tfrac{1}{2}$), which by itself would be ok, but 1 satisfies the δ-ε criterion for all smaller δ as required by the trial definition. This will force us to abandon this definition.

For simplicity we start with the same $\delta = \tfrac{1}{2}$ and try $\varepsilon = \tfrac{3}{2}$ as suggested by the figure. We must determine if

whenever $\qquad |x - 1| < \tfrac{1}{2}$

then $\qquad |y - 1| < \tfrac{3}{2}$

This is demonstrated by the following sequence of relations each of which follows immediately from the previous one.

$|x-1| < \frac{1}{2}$ by assumption

$x \in \left(\frac{1}{2}, \frac{3}{2}\right)$ restate in different notation

$x - \frac{3}{4} \in \left(-\frac{1}{4}, \frac{3}{4}\right)$ shift in the negative direction by $\frac{3}{4}$

$\left|x - \frac{3}{4}\right| < \frac{3}{4}$ largest absolute value in interval is $\frac{3}{4}$

$\left|2x - \frac{3}{2}\right| < \frac{3}{2}$ multiply by 2

$\left|\left(2x - \frac{1}{2}\right) - 1\right| < \frac{3}{2}$ rearrange

$|y - 1| < \frac{3}{2}$ use definition of y

So as stated earlier, we have now shown that

whenever $|x-1| < \frac{1}{2}$

then $|y-1| < \frac{3}{2}$

This means that given $\delta = \frac{1}{2}$ we have found a suitable ε, namely $\varepsilon = \frac{3}{2}$, all in an attempt to show that 1 satisfies the δ-ε criterion in the trial definition of $\lim_{x \to 1} f(x)$. This is not a good indicator because we do not want to succeed with any value for the limit other than $\frac{3}{2}$, but this is just one value of δ and the δ-ε criterion must be satisfied for any δ.

We did start with $\delta = \frac{1}{2}$ so perhaps a smaller value would fail as we desire. But if we were to proceed with a smaller δ, we would find the same unexpected behavior we already found for the limit value $\frac{3}{2}$. We would find

$\varepsilon = \frac{3}{2}$ satisfies the definition for all smaller δ. Thus 1 is also a limit of $\lim_{x\to 1} f(x)$. As Figure 20 visually shows, this makes no sense. We are forced to discard our trial definition.

Conclusion

It should be clear that the problem with the trial definition is that there is nothing to force y to get arbitrarily close to c, i.e., nothing to force ε to get arbitrarily close to 0. We are only able to force x to get arbitrarily close to b in our selection of δ.

Starting from Figure 15 our intention is that as δ gets smaller so does ε, as shown in Figure 16. While the trial definition permits this, it also permits the situation shown in Figure 21 where ε remains unchanged as δ gets smaller.

Day 5 will show how to fix this.

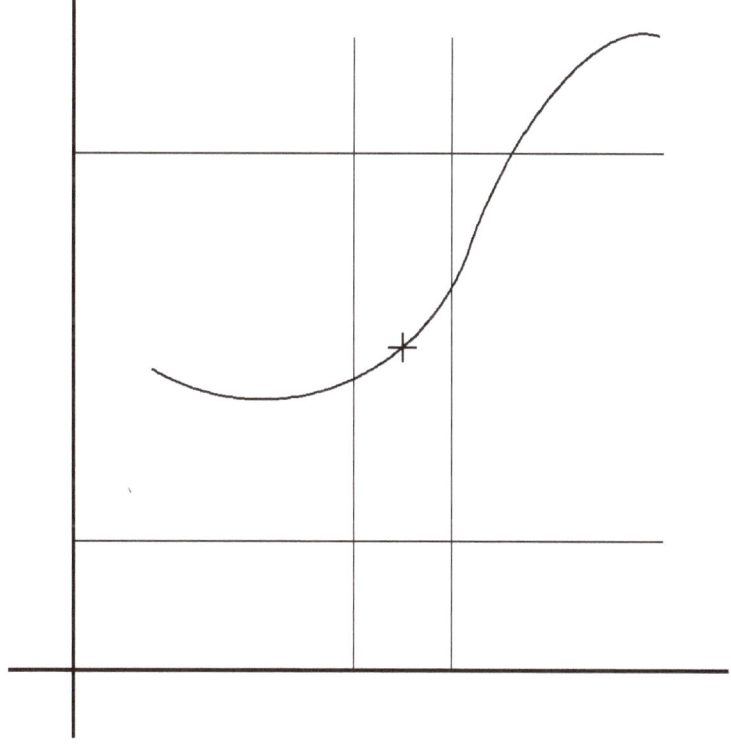

Figure 21. The problem with the trial definition.

41

Day 5: Limits: Rigor That Works

Overview

Since in Day 4 we were able to force x close to b by our selection of δ, this suggests a simple fix. Let us select ε first so that we are able to force y close to c. Only then will we select a suitable δ to define the allowed values of x. As we will see, this change in the order of selection will do exactly what we want. We have arrived at the standard definition of a limit.

Definition. A function $f(x)$ is said to have a *limit* c as $x \to b$, written $\lim_{x \to b} f(x) = c$, if given any number $\varepsilon > 0$ we can find some $\delta > 0$ (which will depend on ε) such that

whenever $\qquad |x - b| < \delta$

then $\qquad |y - c| < \varepsilon$

We now analyze the same function $f(x) = 2x - \frac{1}{2}$ and the value of $\lim_{x \to 1} f(x)$.

Does the Expected Limit Satisfy the Definition of a Limit?

First, we show that $\frac{3}{2}$ satisfies the definition of the limit. As before the inequalities to be satisfied are

whenever $\qquad |x - 1| < \delta$

then $\qquad \left| y - \frac{3}{2} \right| < \varepsilon$

Start with $\varepsilon = 1$. From Figure 22a it is visually clear that this forces $\delta < \frac{1}{2}$. If δ were larger, y would not be constrained to be within the horizontal lines. Figure 22b shows that a smaller delta also works.

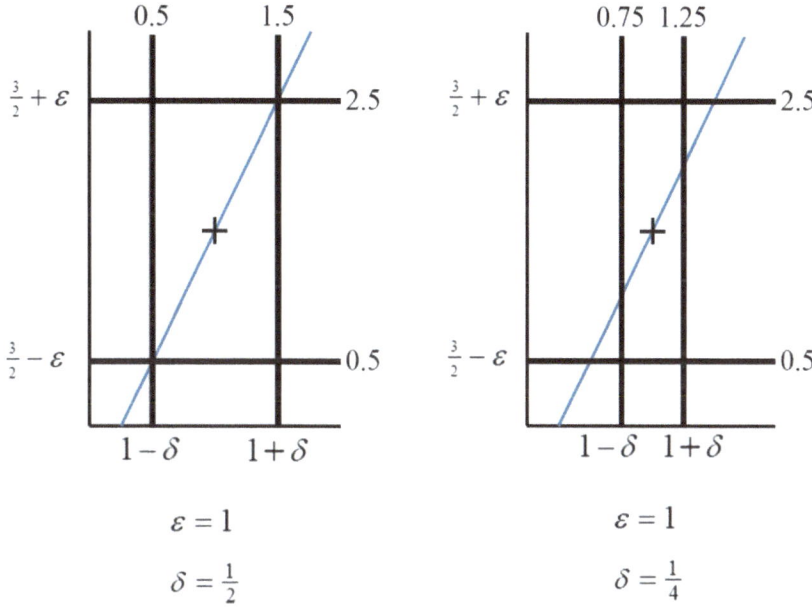

$$\varepsilon = 1 \qquad\qquad\qquad \varepsilon = 1$$
$$\delta = \tfrac{1}{2} \qquad\qquad\qquad \delta = \tfrac{1}{4}$$

Figure 22 a & b.   Initial $\delta - \varepsilon$ diagrams to see if $\lim\limits_{x\to 1} f(x) = \tfrac{3}{2}$.   Choosing $\varepsilon = 1$ forces $\delta \leq \tfrac{1}{2}$. Two possible choices of δ are shown. The second value $\delta = \tfrac{1}{4}$ is randomly chosen as a typical number smaller than $\tfrac{1}{2}$.

This shows an unanticipated bonus in the definition. Picking δ first in the trial definition created a lower bound on ε, but picking ε first in the correct definition creates an upper bound on δ. By forcing y to be close to c we automatically force x to be close to b, which is exactly what we want.

The inequalities still must be proved. Plugging in the values $\varepsilon = 1$ and $\delta = \tfrac{1}{2}$, we must prove that

whenever $\qquad \left| x - 1 \right| < \tfrac{1}{2}$

then $\qquad\quad \left| y - \tfrac{3}{2} \right| < 1$

The following sequence proves this.

$$|x-1| < \tfrac{1}{2} \qquad\qquad \text{by assumption}$$

$$|2x-2| < 1 \qquad\qquad \text{multiply by 2}$$

$$\left|(2x-\tfrac{1}{2})-\tfrac{3}{2}\right| < 1 \qquad\qquad \text{rearrange so we can use definition of } y$$

$$\left|y-\tfrac{3}{2}\right| < 1 \qquad\qquad \text{use definition of } y \text{ aka } f(x) \text{ on page 42}$$

We have now succeeded with $\varepsilon = 1$ by using $\delta = \tfrac{1}{2}$, but we must succeed for all smaller ε too. It turns out that selecting $\delta = \tfrac{1}{2}\varepsilon$ always works as the following sequence shows. The sequence above is just a special case of this sequence. We start by assuming $|x-1| < \tfrac{1}{2}\varepsilon$.

$$|x-1| < \tfrac{1}{2}\varepsilon \qquad\qquad \text{by assumption}$$

$$|2x-2| < \varepsilon \qquad\qquad \text{multiply by 2}$$

$$\left|(2x-\tfrac{1}{2})-\tfrac{3}{2}\right| < \varepsilon \qquad\qquad \text{rearrange}$$

$$\left|y-\tfrac{3}{2}\right| < \varepsilon \qquad\qquad \text{use definition of } y$$

Given any ε we have found a suitable δ, namely $\delta = \tfrac{1}{2}\varepsilon$. This proves that $\tfrac{3}{2}$ satisfies the δ-ε criterion in the definition of the limit.

Does Any Other Number Also Satisfy the Definition?

The last step is to show that no other number also satisfies the definition. We use one particular number to demonstrate how this works in general. Specifically, we again attempt to show that 1 also satisfies the definition of the limit. Again we hope to fail.

Having ε at our disposal, we choose a value smaller than the difference between the real limit $\frac{3}{2}$ and the supposed limit 1. Specifically, we choose $\varepsilon = \frac{1}{4}$ as depicted in Figure 23. In both graphs, when $x \in (1-\delta, 1+\delta)$ there are corresponding values of y that are not within the interval $(1-\varepsilon, 1+\varepsilon)$. These are drawn in red.

Figure 23b shows the δ-ε diagram for $\varepsilon = \frac{1}{4}$ and $\delta = \frac{1}{8}$. It is visually clear from that diagram that for any δ even smaller than $\frac{1}{8}$, meaning the vertical lines are even closer together than shown, that the difference $|y-1|$ between 1 (our presumed limit) and the values of $f(x)$ near $x = 1$ cannot be made smaller than ε.

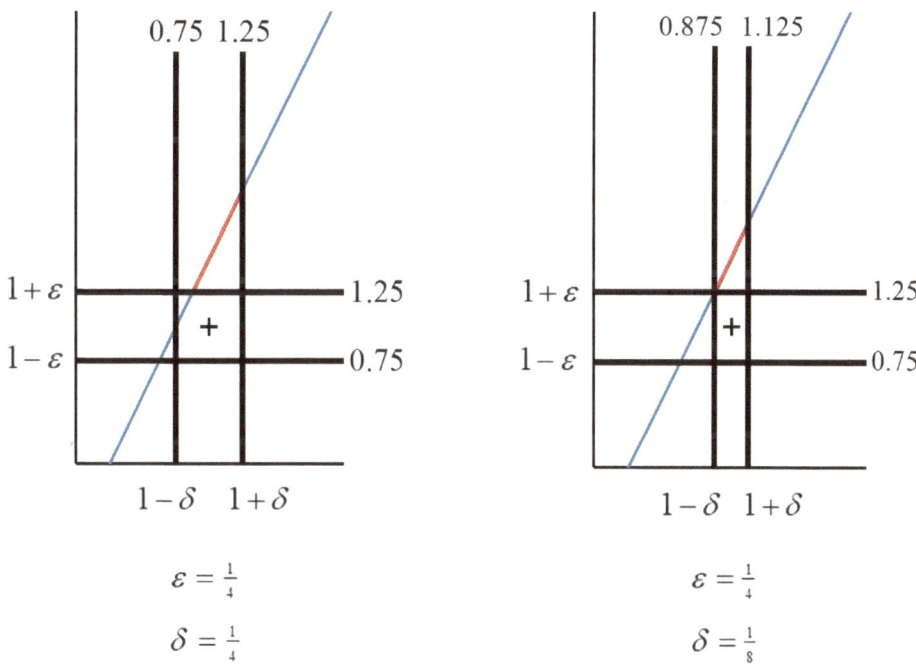

Figure 23 a & b. Initial δ-ε diagrams to see if $\lim f(x) = 1$. The cross marks the point (1,1). Two possible choices of δ are shown, neither of which work.

Again, a diagram is not a proof. Taking any $\delta < \frac{1}{8}$ and starting with $|x-1| < \delta$, the following sequence does prove this.

$|x-1| < \delta < \frac{1}{8}$	by assumption
$|2x-2| < \frac{1}{4}$	multiply by 2
$\left|(2x-\frac{1}{2})-\frac{3}{2}\right| < \frac{1}{4}$	rearrange
$\left|y-\frac{3}{2}\right| < \frac{1}{4}$	use definition of y
$y \in (\frac{5}{4}, \frac{7}{4})$	restate in different notation
$y-1 \in (\frac{1}{4}, \frac{3}{4})$	move in the negative direction by 1
$|y-1| > \frac{1}{4} = \varepsilon$	lower bound of absolute value in the interval $(\frac{1}{4}, \frac{3}{4})$ is $\frac{1}{4}$

Now for 1 to satisfy the definition of $\lim_{x \to 1} f(x)$ for $f(x) = 2x - \frac{1}{2}$ we would need to prove that

$$\text{whenever} \qquad |x-1| < \delta$$

$$\text{then} \qquad |y-1| < \varepsilon$$

But the sequence above proves exactly the opposite.

By itself this is not enough to conclude that 1 is not a limit, for this negative result is so far only for $\varepsilon = \frac{1}{4}$. But the same result occurs for any ε by choosing any $\delta < \frac{1}{2}\varepsilon$. Therefore, we cannot find suitable δ given any ε and must conclude that 1 does not satisfy the definition of $\lim_{x \to 1} f(x)$.

This same argument also works for any number other than 1 except for $\frac{3}{2}$. We conclude that $\frac{3}{2}$ satisfies the definition of the limit and that no other number does.

This argument is easily modified to prove that <u>any</u> function has at most one limit, a necessary characteristic of a meaningful definition of a limit.

Conclusion

While we have only used linear functions so far, it should be clear that this definition of a limit does everything we want it to do. As a result, this definition of a limit is universally adopted.

It was historically a surprise that this new concept, the most basic definition in establishing a rigorous formulation of calculus, is based on inequalities. Inequalities are not where one expects to find contributions to great mathematical truth or insight. This is a contributing and perhaps a major reason it took so long to discover.

Limits of Non-linear Functions

This section is optional.

The purpose of this short section is to show a computation of a limit with a non-linear function, for linear functions are a special case in which everything works out almost trivially.

Admittedly we will take the simplest non-linear function, $y = x^2$, but it does show some of the practical complications that arise in non-linear cases. We will compute $\lim_{x \to 2} x^2$. We expect this limit to be 4, the value of y when $x = 2$.

To prove 4 is the limit according to the definition we must demonstrate that we can find δ, depending on ε, such that whenever $\quad |x-2| < \delta$

then $\quad |y-4| < \varepsilon$

We are really interested in small values of ε. We restrict ourselves to $\varepsilon < 4$ for reasons relating to the non-linearity of this particular function and which will become clear below. This gives us the following two-step sequence whose conclusion we will use shortly.

$\varepsilon < 4 \qquad\qquad$ by choice as stated above

$\dfrac{\varepsilon^2}{16} < \dfrac{\varepsilon}{4} \qquad\qquad$ multiplying both sides by $\dfrac{\varepsilon}{16}$

Now for any $\varepsilon < 4$ we pick $\delta = \dfrac{\varepsilon}{4}$ and assume $|x-2| < \delta$. In order to prove 4 is the limit we must prove that the δ-ε criterion in the definition of the limit is satisfied by proving $|y-4| < \varepsilon$.

$|x-2| < \delta \qquad\qquad$ by assumption

$x \in (2-\delta, 2+\delta) \qquad$ restate in different notation

$y \in \left((2-\delta)^2, (2+\delta)^2 \right) \qquad$ take squares ($y = x^2$)

$y \in \left(4-2\delta+\delta^2, \ 4+2\delta+\delta^2 \right) \qquad$ expand the squares

$y-4 \in \left(-2\delta+\delta^2, \ 2\delta+\delta^2 \right) \qquad$ shift by 4

$|y-4| < 2\delta+\delta^2 \qquad$ largest absolute value in interval is $2\delta+\delta^2$

$|y-4| < 2\dfrac{\varepsilon}{4} + \dfrac{\varepsilon^2}{16} \qquad$ we chose $\delta = \dfrac{\varepsilon}{4}$

$$|y-4| < 2\frac{\varepsilon}{4} + \frac{\varepsilon}{4} \qquad \text{use } \frac{\varepsilon^2}{16} < \frac{\varepsilon}{4} \text{ from above}$$

$$|y-4| < \frac{3\varepsilon}{4} < \varepsilon \qquad \text{simplify}$$

This is what we had to prove given our choice of δ. Having found a suitable δ for any ε we have, by the definition of a limit, proved that $\lim_{x\to 2} x^2 = 4$.

Limit Theorems

Here are some basic theorems about limits that we state without proof, the proofs being straightforward but detailed exercises in δ-ε manipulation. The limit of a sum, difference, product or quotient equals the sum, difference, product or quotient of the limits, provided all limits on the right exist and we do not divide by 0. The fifth theorem says that constants can be taken outside a limit.

$$\lim_{x\to a} \big(f(x) + g(x)\big) \;=\; \lim_{x\to a} f(x) \;+\; \lim_{x\to a} g(x)$$

$$\lim_{x\to a} \big(f(x) - g(x)\big) \;=\; \lim_{x\to a} f(x) \;-\; \lim_{x\to a} g(x)$$

$$\lim_{x\to a} f(x)g(x) \;=\; \Big(\lim_{x\to a} f(x)\Big)\Big(\lim_{x\to a} g(x)\Big)$$

$$\lim_{x\to a} \frac{f(x)}{g(x)} \;=\; \frac{\lim_{x\to a} f(x)}{\lim_{x\to a} g(x)} \qquad \text{provided } \lim_{x\to a} g(x) \neq 0$$

$$\lim_{x\to a} cf(x) \;=\; c \lim_{x\to a} f(x) \qquad \text{for constant } c$$

In Day 3 we discussed variables traveling toward their limiting values while in Day 5 the situation is static, no traveling is involved. This is a difference between Cauchy's original conception, which is maintained in Day 3 for its easy intuition, and Weierstrass' modern formulation, which dispenses with the notion of traveling.

Problems

1. Let $f(x) = \frac{1}{x}$ as shown. It is visually clear that $\lim\limits_{x \to 0} \frac{1}{x}$ does not exist. Prove it.

   Hint: Assume the limit does exist with value c, i.e., $\lim\limits_{x \to 0} \frac{1}{x} = c$. Assume $c \geq 0$, the proof for $c < 0$ being all but identical. Pick a value for ε. For any δ smaller than some threshold show that the δ-ε criterion is violated for any positive x within δ of 0.

   Second hint: See footnote if desired.[4]

2. Let $f(x) = \frac{1}{x}$ and $g(x) = \frac{1}{x-1}$. Can the first limit theorem (for sums) be applied when $b = 0$? $b = 1$? $b = 2$? Why or why not?

3. Let $f(x) = \frac{1}{x}$ and $g(x) = \frac{1}{x}$. Can the second limit theorem (for differences) be applied when $b = 0$? Why or why not?

   Does the limit on the left exist when $b = 0$? If so, what is its value?

_____

[4] Second hint: select $\delta < \frac{1}{c+\varepsilon}$

51

Day 6: Continuity: The Key to Everything

Continuity Finally Defined

Definition. A function $f(x)$ is *continuous* at a point a if the following is true.

$$\lim_{x \to a} f(x) = f(a)$$

In order for this equation to be valid all of the following must be true.

- the limit of the function as $x \to a$ must exist so that the left side of the equation exists,

- the function must be defined at a so that the right side of the equation exists, and

- the limit of the function as $x \to a$ must equal the value of the function at a.

This is the Equation of Continuity. Understand it and memorize it.

A function is said to be continuous *in an interval* (p,q) if the Equation of Continuity holds at every point in the interval.

Important Continuous Functions

Constant functions

Suppose $f(x) = c$ for all x. Pick any value of x, say $x = a$. Then $f(a) = c$. We now prove that f is continuous at a. This should be obvious, but it is useful to see how the definition of continuity is actually used to prove a function is continuous.

The right side of the Equation of Continuity has the value c. We show that the limit on the left side also has the value c, which implicitly shows that the limit exists.

Pick any positive ε however small. Normally δ would be selected as a function of ε and would itself be small. However, for a constant function we can pick any δ at all, so just to make that point clear we arbitrarily select $\delta = 17$. Is it the case that for all x such that $|x-a| < 17$ that we have $|f(x)-c| < \varepsilon$? Since $f(x) = c$ the latter inequality reduces to determining if $|c-c| < \varepsilon$, i.e., if $0 < \varepsilon$ everywhere in the interval $(a-17, a+17)$. This is certainly true. This means that for any ε we have found an appropriate δ, so the limit exists and has the value c. We conclude that the Equation of Continuity is satisfied and that constant functions are continuous.

Identity function

Suppose $f(x) = x$. Pick any value of x, say $x = a$. Then $f(a) = a$. We now prove that f is continuous at a.

The right side of the Equation of Continuity has the value a. We show that the limit on the left side also has the value a, which implicitly shows that the limit exists.

Pick any positive ε however small. Selecting δ is easy for the identity function, we just select $\delta = \varepsilon$. Is it the case that for all x such that $|x-a| < \delta$ that $|f(x)-a| < \varepsilon$? Since $x = f(x)$ and $\delta = \varepsilon$ these two inequalities are identical, so when the first is true so is the second. This means that for any ε we have found an appropriate δ, so the limit exists and has the value a. We conclude that the Equation of Continuity is satisfied and that the identity function is continuous.

Combinations of continuous functions

Suppose $f(x)$ and $g(x)$ are continuous at point a. Simple theorems state that a function h given by their sum, difference, product or quotient $(f+g, f-g, fg, f/g)$ are all continuous at point a, save for the quotient if $g(a) = 0$. The proof of each theorem proceeds by using the Equation of Continuity for f and g to show it is also satisfied by h. The Limit Theorems listed on page 49 are also used.

Here is the proof that the sum of continuous functions is continuous. Let $h(x) = f(x) + g(x)$. We must prove that $\lim_{x \to a} h(x) = h(a)$.

$$\lim_{x \to a} h(x) = \lim_{x \to a} (f(x) + g(x)) \qquad \text{by definition of } h$$

$$= \lim_{x \to a} f(x) + \lim_{x \to a} g(x) \qquad \text{by Limit Theorem for sums}$$

$$= f(a) + g(a) \qquad \text{by continuity of } f \text{ and } g$$

$$= h(a) \qquad \text{by definition of } h$$

Polynomials

Consider a randomly chosen polynomial such as

$$f(x) = 5x^3 - 3x^2 + 13x + 6$$

This polynomial is constructed from x (the identity function) and from constant functions by taking the products of the various constants with multiple instances of x, then adding the resulting terms. All that is needed to prove this or any polynomial is continuous are the three preceding results: constants are continuous, the identity function is continuous and simple combinations of continuous functions are continuous.

The Equation of Continuity is not explicitly used in this proof because the proof uses the immediately preceding result about combinations of continuous functions instead.

Not explicitly using the Equation of Continuity will be of major significance in Day 7.

Strange Discontinuous Behavior

This section is optional and may be skipped, especially by students who find it too abstract.

A rational number is the ratio of two integers. $\frac{3}{2}$ is an example. Conversely, an irrational number is one that cannot be expressed as a ratio of integers.

This criterion for rationality is whether a number can somehow be expressed as a ratio of integers, not whether that is the only way it can be expressed. For example, $\frac{\pi}{2\pi}$ is not expressed as the ratio of integers but it can be expressed as such in the form $\frac{1}{2}$ and so it is rational. $\sqrt{2}$ cannot be so expressed and is irrational.

The rational and irrational numbers are intimately mixed on the number line. Every interval, however small, contains an infinite number of both rationals and irrationals.

Continuity is a strong constraint on the behavior of a function. Without knowing that a function is continuous throughout some interval, a function's continuity properties can be strange. The three example functions defined here have very different continuity properties despite having definitions that may not seem to justify the extent of the differences.

Example 1

$$f(x) = 1 \qquad \text{if } x \text{ is rational}$$
$$f(x) = 0 \qquad \text{if } x \text{ is irrational}$$

This function is continuous nowhere. Its graph is shown in Figure 24. The two lines are not really lines, for each has an infinite number of points missing that make up the other apparent line.

Figure 24 The function of example 1.

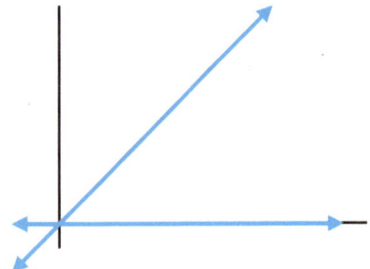

Figure 25 The function of example 2.

Example 2

$$f(x) = x \qquad \text{if } x \text{ is rational}$$
$$f(x) = 0 \qquad \text{if } x \text{ is irrational}$$

This function is continuous at only one point, the point $x = 0$. Its graph is shown in Figure 25.

Example 3

$$f(x) = \frac{1}{b} \qquad \text{if } x \text{ is a rational number } \tfrac{a}{b} \text{ expressed in lowest}$$

terms with the denominator b being positive, e.g., if $x = \tfrac{3}{2}$ then $f(x) = \tfrac{1}{2}$

$$f(x) = 0 \qquad \text{if } x \text{ is irrational}$$

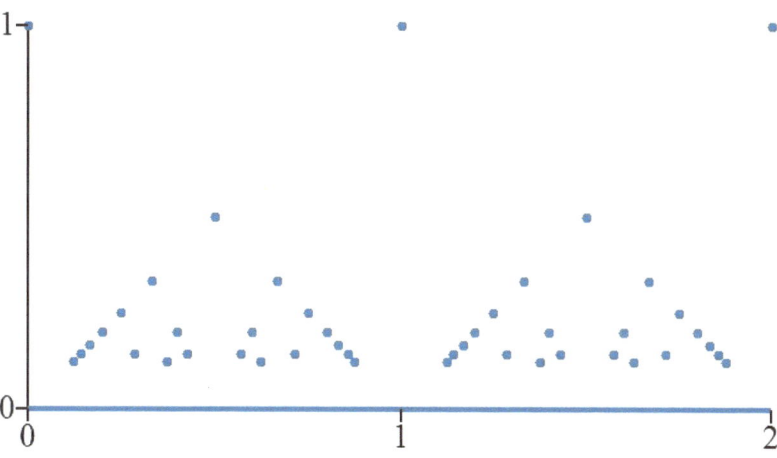

Figure 26 The function of example 3 with all rational
numbers expressible with a denominator of 8 or less.

While not obvious, this function is continuous when x is irrational and discontinuous when x is rational, an intimate mixing of the points where the function is continuous and discontinuous. This function cannot be completely graphed but is partially graphed in Figure 26. The apparent line along the x-axis consists of the irrational points only. Values of selected rational points are shown. The remaining rational points lie even closer to the x axis. The larger the denominator of a rational number the closer it is to the x axis.

Another Example of Using the Definition of Continuity

This section continues from the previous section and is also optional.

To get you more practiced in the use of limits, we now prove that the function in example 1 is discontinuous at the irrational numbers.

Pick an irrational number a. In order for f to be continuous at a the limit on the left side of the Equation of Continuity would have to exist and have the right value. Since $f(a) = 0$ the right value is 0.

Choose $\varepsilon < 1$ so that the distance between the two possible function values (which are 0 and 1) is larger than ε. To be specific we choose $\varepsilon = 0.5$. If it were true that $\lim_{x \to a} f(x)$ existed and that $\lim_{x \to a} f(x) = 0$, then for our chosen value $\varepsilon = 0.5$ we would be able to find some sufficiently small δ such that

whenever $\quad |x-a|<\delta$

then $\quad |y-0|<0.5 \quad$ or $\quad |y|<0.5$

But regardless of how small δ is selected, there are rational numbers r closer to a than the distance δ, i.e., $|r-a|<\delta$. For these r we have $y=1$, which contradicts the requirement that $|y|<0.5$.

We conclude that the limit on the left side of the Equation of Continuity does not exist with the value 0 (in fact, it does not exist with any value). The Equation of Continuity is not satisfied and the function is discontinuous at the irrational points.

The proof that it is also discontinuous at rational points is the same proof with some numbers changed.

Problems

These problems are about the optional sections of this Day.

1. Prove that the function in example 2 is continuous at 0. Hint: For any ε select $\delta = \varepsilon$. Consider nearby rationals and irrationals separately. Note that when x is rational $f(x)$ is the identity function and when x is irrational it is a constant function.

2. Prove that the function in example 3 is discontinuous at rational points $\frac{a}{b}$. Hint: $f(\frac{a}{b}) = \frac{1}{b}$ so choose $\varepsilon < \frac{1}{b}$. Use the fact that there are irrational numbers arbitrarily close to $\frac{a}{b}$.

Day 7: Derivatives: Putting It All Together

Using Continuity in Reverse

The Equation of Continuity is used to prove a function $f(x)$ is continuous, but there is another use of the Equation of Continuity. We can often show $f(x)$ is continuous based on knowledge of other functions already known to be continuous. The result discussed in Day 6 that all polynomials are continuous is one example. Review that brief section (page 54) if you do not remember it.

Once we know $f(x)$ is continuous by other means, we know the Equation of Continuity is true and it is available for our use. If we then have to evaluate $\lim_{x \to b} f(x)$ we simply evaluate $f(b)$ instead. This is using continuity in reverse. You will commonly do this when computing derivatives and thereby avoid δ-ε manipulation.

Final Flashback - The Function That Started It All

Before tackling the derivative, it is time to show the power of the tools we have developed. With the limit as a new tool and an understanding of continuity, it is now an effortless task to rigorously show that the function in Figure 8 when evaluated at $x = 2$ would be a nice function if its value were 3 instead of 0.

From the function definition we have

$$\lim_{x \to 2} f(x) = \lim_{x \to 2} x + 1 = 3$$

Because the limit does not use the value of $f(x)$ at the point $x = 2$, it is irrelevant that $f(x)$ is not defined by $x + 1$ at $x = 2$. We also used the facts that $x + 1$ is a polynomial and that polynomials are continuous. This allows us to evaluate the limit by using continuity in reverse and so just plug in $x = 2$ in the formula $x + 1$. So $f(x)$ would be a continuous function if $f(2) = 3$ instead of $f(2) = 0$.

Derivatives

You will shortly see or perhaps have already seen in your course of calculus study the definition of a derivative. The intention is that the derivative of $y = f(x)$ is the slope of the curve of $f(x)$, which is itself defined as the slope of the line that is tangent to the curve.

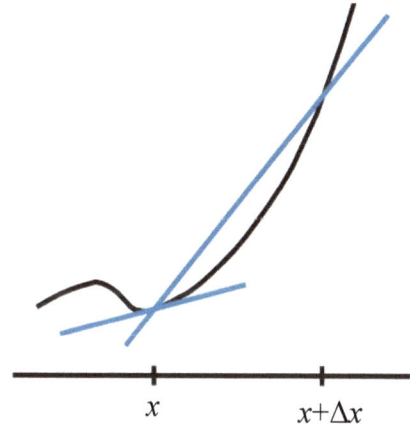

Figure 27 A secant (see text for definition) and a tangent, both based at point x.

Δ is the upper case form of the Greek letter delta whereas δ is the lower case form. "Δx" (pronounced "delta x") is used as a single symbol to denote an increment in x, usually a small increment. It may be positive or negative.

From algebra we know that the slope of a straight line is

$$s = \frac{\Delta y}{\Delta x}$$

Knowing how to compute $f(x)$ means that for any x we can compute the corresponding y on the curve by the equation $y = f(x)$. Unfortunately, as shown in Figure 27, we cannot use this to get the slope of a tangent because the tangent intersects the curve at only one point whereas we need two points on a straight line to compute Δy.

So algebra provides no method to compute the slope of the tangent line. What algebra does provide is a method to compute the slope of a secant, a secant being any straight line that intersects a curve in two or more points.

Consider the secant in Figure 28. It intersects the graph of f at x and at $x+\Delta x$ with corresponding y values $f(x)$ and $f(x+\Delta x)$, respectively. Given these two points we have

$$\Delta y = f(x+\Delta x) - f(x)$$

Keeping the left point fixed at $(x, f(x))$ and moving only the right point $(x+\Delta x, f(x+\Delta x))$ along the curve by varying Δx, we get the slope s of the secant as a function of Δx. Note well that once we have picked x, it is a constant. s is then not a function of x, it is only a function of Δx:

$$s(\Delta x) = \frac{f(x+\Delta x) - f(x)}{\Delta x}$$

This formula is simple to evaluate for any secant based at x, i.e., for any $\Delta x \neq 0$.

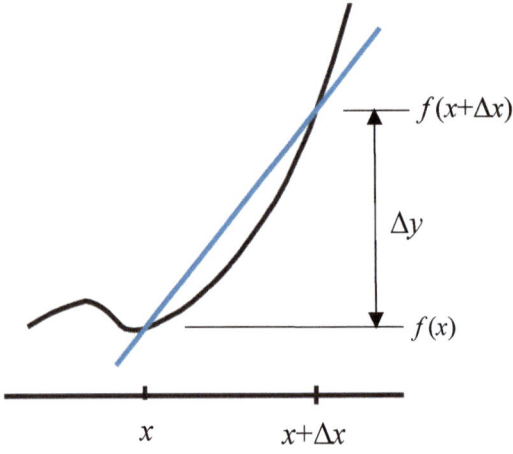

Figure 28 Computing the slope of a secant.

But we are only interested in the slope when $\Delta x = 0$, which is when the secant becomes the tangent, for it is tangents whose slopes give us the slope of the curve at each point. So the function s tells us nothing of interest except that, if it were defined at $\Delta x = 0$, that would tell us the derivative (whose value is the slope) at the current value of x. In conclusion, for each value of x we want the value of $s(\Delta x)$ at the one and only point ($\Delta x = 0$) where it is not defined.

Algebra cannot save this situation but limits can. We choose to fill in the missing point in the graph (not pictured) of $s(\Delta x)$ by defining $s(0)$ so that s is continuous at 0. We use the Equation of Continuity to define s at $\Delta x = 0$ as follows.

$$s(0) = \lim_{\Delta x \to 0} s(\Delta x)$$

This assumes the limit exists and we leave $s(0)$ undefined if the limit does not exist. Using the formula for $s(\Delta x)$ gives

$$s(0) = \lim_{\Delta x \to 0} \frac{f(x + \Delta x) - f(x)}{\Delta x}$$

Of course, $s(0)$ is just the slope of the tangent, i.e., the derivative of $f(x)$, at the point x and the usual way of writing this is

$$\frac{dy}{dx} = \lim_{\Delta x \to 0} \frac{f(x + \Delta x) - f(x)}{\Delta x}$$

where $\frac{dy}{dx}$ is the standard notation for a derivative. It is read "d y over d x". Having defined the derivative, calculus takes off from this point.

Practical computation of derivatives

Usually we can skip the laborious δ-ε process when computing the limit that is in the definition of a derivative. A simple example illustrates this best. Suppose we want to compute the derivative of $y = x^2$. We have

$$\frac{dy}{dx} = \lim_{\Delta x \to 0} \frac{(x + \Delta x)^2 - x^2}{\Delta x} = \lim_{\Delta x \to 0} \frac{2x\Delta x + \Delta x^2}{\Delta x} = \lim_{\Delta x \to 0} 2x + \Delta x = 2x$$

Each equality in this sequence will seem obvious if you have already been exposed to derivatives, but it is still well worth examining them.

The first equality is just an application of the definition of the derivative.

The second equality comes from applying the basic rules of algebra to simplify the numerator.

The third equality deserves special attention because there is a subtlety here beyond the obvious further simplification via algebra. The formulas $\frac{2x\Delta x + \Delta x^2}{\Delta x}$ and $2x + \Delta x$ do not define the same function, for the first is undefined at $\Delta x = 0$ while the second is defined there. Taking the limit of a different function requires a justification. The justification is based on the fact that the new function $2x + \Delta x$ is equal to the original function $\frac{2x\Delta x + \Delta x^2}{\Delta x}$ wherever the original function is defined, i.e., for all $\Delta x \neq 0$. Because the limit only uses function values at points other than $\Delta x = 0$, we are justified in canceling Δx inside the limit and using a function defined at the one additional point $\Delta x = 0$. We are now set up for the fourth equality, which will use this additional point.

For the fourth equality, remember that x is being held constant in this limit process and Δx is the variable. The expression $2x + \Delta x$ is a linear polynomial in Δx and we know all polynomials are continuous. Therefore, we can use continuity in reverse. By the Equation of Continuity, we can evaluate the limit just by plugging in $\Delta x = 0$ which gives us $2x$.

This example shows how you will almost always compute derivatives. When you arrive at an expression in which it looks like you can just plug in 0 you should go ahead and do so. However, if you cannot cancel Δx or if you can cancel but do not get a function that is known to be continuous, it is necessary to go through the δ-ε process to evaluate the limit.

Just to provide one such example, an important case of not being able to use continuity in reverse occurs when differentiating the trigonometric function sin x. This comes down to evaluating the following two limits via the δ-ε process.

$$\lim_{\Delta x \to 0} \frac{\sin \Delta x}{\Delta x} \qquad\qquad \lim_{\Delta x \to 0} \frac{(\cos \Delta x) - 1}{\Delta x}$$

It is an important theorem with a non-trivial δ-ε proof that

$$\lim_{\Delta x \to 0} \frac{\sin \Delta x}{\Delta x} = 1 \qquad\qquad \lim_{\Delta x \to 0} \frac{(\cos \Delta x) - 1}{\Delta x} = 0$$

Problems

1. Compute the derivative of $y = x^3$.

You do not need to have studied trigonometry to do the following problems. Any required trigonometric knowledge is supplied in the problem hints.

2. Show that differentiating $\sin x$ reduces to evaluating the two limits stated in the text, $\lim_{\Delta x \to 0} \dfrac{\sin \Delta x}{\Delta x}$ and $\lim_{\Delta x \to 0} \dfrac{(\cos \Delta x) - 1}{\Delta x}$. Hint: Write the expression for the derivative using its definition, then use the trigonometric identity $\sin(a+b) = \sin a \cos b + \sin b \cos a$. You will also need the limit theorems listed on page 49. Note that with respect to Δx both $\sin x$ and $\cos x$ are constants.

3. Show that differentiating $\cos x$ also reduces to evaluating the same two limits.   Hint: Use   the   trigonometric   identity $\cos(a+b) = \cos a \cos b - \sin a \sin b$.

Derivative Notation and Differentials

Modern mathematics predominantly uses Leibniz's notation for derivatives, which is $\frac{dy}{dx}$. This shows how Leibniz thought of derivatives as ratios of mysterious infinitesimal quantities as discussed in Day 1.

The standard modern definition does not define derivatives as ratios but rather as limits. The definition given as the limit of a certain expression is holistic, meaning that the definition defines the entire object $\frac{dy}{dx}$ without regard to any component parts that may or may not exist.

Certainly the notation $\frac{dy}{dx}$ is suggestive that there are component parts denoted dy and dx whose ratio we are taking, but this suggestion is misleading and false. $\frac{dy}{dx}$ has no component parts, for the limit of the ratio in the definition of the derivative

$$\lim_{\Delta x \to 0} \frac{f(x + \Delta x) - f(x)}{\Delta x}$$

is not itself a ratio of limits, despite the limit theorems at the end of Day 5. This is because taking the separate limits of numerator and denominator gives $\frac{0}{0}$. The limit theorem for a quotient explicitly states it does not apply when the limit of the denominator is 0.

The only way entities labeled dy and dx can exist on their own is if we choose to independently define them, which in fact we do choose to do. Despite intuition we do not define them as infinitesimals because infinitesimals do not exist in the standard formulation of calculus. Instead we define them as ordinary numbers in such a way that their ratio gives the same value as the derivative $\left(\frac{dy}{dx}\right)$. How this is done may seem like a cheap trick but it works.

Whatever value is freely chosen for dx (usually a small number), dy is defined by

$$dy = \left(\frac{dy}{dx} \right) dx$$

where $\left(\frac{dy}{dx} \right)$ is not a ratio but is still the derivative as defined by a limit. One cannot cancel what appear to be two instances of dx on the right side of this equation, for there is only one such instance.

Dividing both sides of the equation by dx gives

$$\frac{dy}{dx} = \left(\frac{dy}{dx} \right)$$

where the left side is a ratio and the right side is a limit. Take good note of this, that the difference between the left and right sides is not just a pair of parentheses but rather is a matter of fundamental definition.

Because of this equation one is free to interpret $\frac{dy}{dx}$ in any context either as a ratio or as a limit. Of course, the actual computation of $\frac{dy}{dx}$ requires the evaluation of the limit.

The independent definitions of dy and dx justify, to the extent one may feel it needs justification, our use of the Leibniz notation even though we have not chosen to base calculus on infinitesimals. Not being infinitesimals in this limit-based formulation of calculus but rather just ordinary numbers, dy and dx are each given the name *differential* and, as just discussed, the ratio of these two differentials is the derivative. Selecting a different value of dx yields a different value of dy but the ratio is still the same.

It is still the case that dy and dx are not component parts of the derivative $\frac{dy}{dx}$ even though we defined them to have the right ratio. This is because $\frac{dy}{dx}$ is *defined* as a certain limit even though it happens to also equal that ratio.

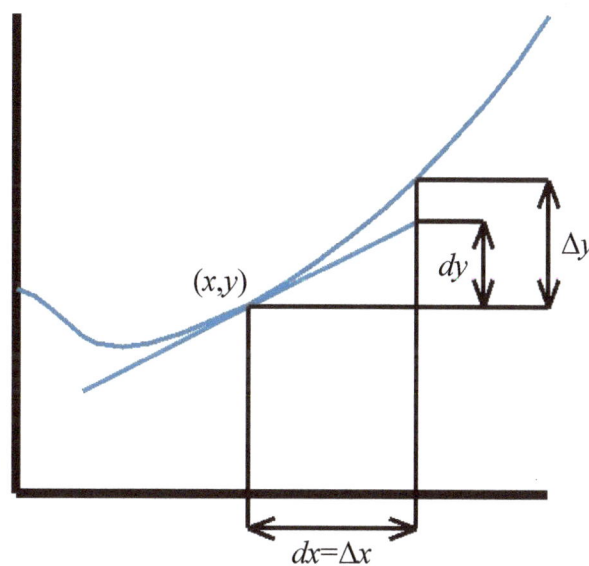

Figure 29 The relationship between dy and Δy.

Nevertheless, many people choose to think of *dy* and *dx* as component parts as though they were infinitesimals simply because it is conceptually easier. If you choose to do this, just be sure to remember that *dy* and *dx* are not actually infinitesimals (which are not defined) but are ordinary numbers. Of course, some people actually do think of them as infinitesimals anyway, for that is both simple and intuitive, but it prevents using differentials as described next.

The practical use of differentials is shown in Figure 29. Take $f(x)$ to be a function easily computable at *x* but difficult to compute at nearby points. It is desired to compute $f(x+\Delta x)$, which has the value *y+Δy*. Because we know this is hard to compute, we settle for an easily computable approximation. We already know how to compute y by $y = f(x)$, so it only remains to compute an approximation to Δy, the actual change in $f(x)$ over the range Δx.

We choose to set $dx = \Delta x$ and then compute dy by multiplying dx by the derivative evaluated at the point x. As can be seen in Figure 29, for small values of Δx we have $dy \approx \Delta y$.

dy is an easily computable, good approximation to Δy.

dy itself is the amount $f(x)$ would have changed if it were the tangent line instead of being a curve.

Problems

1. Let $y = x^2$. Then $\frac{dy}{dx} = 2x$. At $x = 1$ we have $y = 1$ and $\frac{dy}{dx} = 2$. With 1 as the starting point, use differentials to compute an approximation to y at $x = 1.001$. What is the exact value?

2. Round all results in this problem to four digits after the decimal point.

   Let $y = x^3$. Then $\frac{dy}{dx} = 3x^2$ by problem 1 on page 66.

   What is the exact value of y when $x = 2.09$?

   Use differentials to compute an approximation to y at $x = 2.09$ using $x = 2$ as the starting point, i.e., $dx = 0.09$. What is the error in the approximation?

   Use differentials to compute an approximation to y at $x = 2.09$ using $x = 2.1$ as the starting point, i.e., $dx = -0.01$. What is the error in the approximation?

   Graph the function and the two tangent lines over the range [2, 2.1].

3. This problem is for readers who have studied trigonometry. The function $\sin x$ is easy to compute at 0 but hard to compute at $1°$. Using $x = 0$ as the starting point, compute the differential approximation to $\sin 1°$ to six digits after the decimal point. Note that the derivative of $\sin x$ is $\cos x$. Remember to convert degrees to radians. Use a calculator to compute the exact value.

A Look Ahead

This is an optional look ahead to limits as they are used in the second part of calculus.

It was stated in Day 1 in the section "What is Calculus?" that calculus consists of two parts. The second part is called *integral calculus* and deals with computing areas bounded by curves. For example, the formula πr^2 for the area of a circle is typically derived from integral calculus.[5]

In order to compute such areas an extension of the limit definition is required. The extension is illustrated by the sum S of the following infinite series.

$$S = 1 + \tfrac{1}{2} + \tfrac{1}{4} + \tfrac{1}{8} + \tfrac{1}{16} + \ldots$$

It took mathematicians centuries to discover the right way to define the sum of an infinite series. One reason was that they lacked the limit concept and, in particular, lacked the extension we are about to present.

Let S_n be the sum of the first n terms of the series. The first four values are as follows.

$$S_1 = 1$$

$$S_2 = 1 + \tfrac{1}{2} = \tfrac{3}{2}$$

$$S_3 = 1 + \tfrac{1}{2} + \tfrac{1}{4} = \tfrac{7}{4}$$

$$S_4 = 1 + \tfrac{1}{2} + \tfrac{1}{4} + \tfrac{1}{8} = \tfrac{15}{8}$$

---

[5] The formula for the area of a circle has an elementary geometric derivation that was known to the Greeks, but in general integral calculus is required to compute areas.

As n increases these partial sums get closer and closer to 2. In fact, they get arbitrarily close to 2, a fact made plainer by writing the partial sums as follows.

$$S_1 = 1 = 2-1$$

$$S_2 = \frac{3}{2} = 2-\frac{1}{2}$$

$$S_3 = \frac{7}{4} = 2-\frac{1}{4}$$

$$S_4 = \frac{15}{8} = 2-\frac{1}{8}$$

The sum of the *series* is defined as the limiting value of the *sequence* of partial sums:

$$S_1; \quad S_2; \quad S_3; \quad S_4 \quad ...$$

We would like to write $\lim_{n \to \infty} S_n = 2$ but the standard definition of a limit does not allow this, for it is impossible for n to get within any δ of ∞. Thus the standard definition of a limit in which x (or in this case n, which is always an integer) approaches a finite number does not allow for the existence of a limiting value as $n \to \infty$. We must define the new notation $\lim_{n \to \infty}$ so that it does what we want.

Definition: The infinite sequence S_n has a limit S, written $\lim_{n \to \infty} S_n = S$, if given any number $\varepsilon > 0$ we can find some N (which will depend on ε) such that

whenever $\qquad n > N$

then $\qquad |S_n - S| < \varepsilon$

As ε gets smaller the number N typically gets larger without bound.

We now prove that the infinite series above sums to 2. For the partial sums S_n, all of which have a finite number of terms and so are well defined, we have

$$S_n = 2 - \frac{1}{2^{n-1}}$$

For any $\varepsilon > 0$ select N large enough so that $\frac{1}{2^{N-1}} < \varepsilon$ (this can always be done). Then for $n > N$ we have

$$|S_n - 2| = \left|\left(2 - \frac{1}{2^{n-1}}\right) - 2\right| = \frac{1}{2^{n-1}} < \frac{1}{2^{N-1}} < \varepsilon$$

This proves that the infinite series sums to $S = 2$.

Integral calculus does not compute an area as the sum of an infinite series, but its computation does reduce to finding the limiting value of an infinite sequence. Each number in the sequence is a better approximation to the area than the previous number, the error in the approximations going to 0 as n increases without bound, i.e., as n "approaches" infinity.

Newton and Leibniz were not the first to correctly compute slopes and areas. The generation of mathematicians before them had derived results in both these fields. But Newton and Leibniz were the ones who made use of what is now called the Fundamental Theorem of Calculus. This vital theorem (arguably the most important theorem in mathematics) connects differential calculus and integral calculus into a single topic that we now simply call calculus. It is this theorem that provided the foundation that allowed calculus to be developed into the rich and powerful branch of mathematics that it is today, a development that both inventors embarked on when they realized the significance of this recently (at that time) discovered theorem.

This is the end of our discussion of limits and continuity. You may find it useful to refer back to Days 5-7 as you continue learning calculus. You may also develop a new appreciation for the historical notes as laid out mostly in Day 1. In any event, best wishes for your mathematical studies!

Answers to Problems

Page 13

1. This is a function. Each value of x is evaluated by one or the other of the expressions for y but not by both of them, so there is only one value of y for each x.

2. This is not a function. For x in the interval $[0,1]$ both expressions for y are evaluated. For all such values of x except $x = 0$ the two expressions give different values, so this is not a function.

3. This is a function. The number 0 is the one value of x for which both expressions are evaluated for y. Both expressions give the same value 0, so for each x there is at most one y.

Page 17

1. When $m \neq 0$ we can set $y = 0$ and solve for x, dividing by m to get $x = \frac{-b}{m}$. This is the only value we obtain, so the number of roots is 1.

2. When $m = 0$ we cannot solve for x as in the previous problem. Instead we set m to 0 to obtain the new equation $y = b$. x does not appear in this equation, so its value is irrelevant. The value of b is what matters. If $b = 0$ then $y = 0$ for all x, which means there are an infinite number of roots. If $b \neq 0$ then $y \neq 0$ for all x, which means there are no roots. Combining these results with the answer to the previous problem tells us that the number of roots of a linear equation is 0, 1 or ∞.

3. The function has two roots. The roots are −1 and 2. Because linear functions cannot have exactly two roots (by the previous problem) this function is not linear.

Page 21

1. The point is that it does not matter what the function looks like outside the interval [1,3]. Here is one possibility. The squiggles are a face in profile turned 90°.

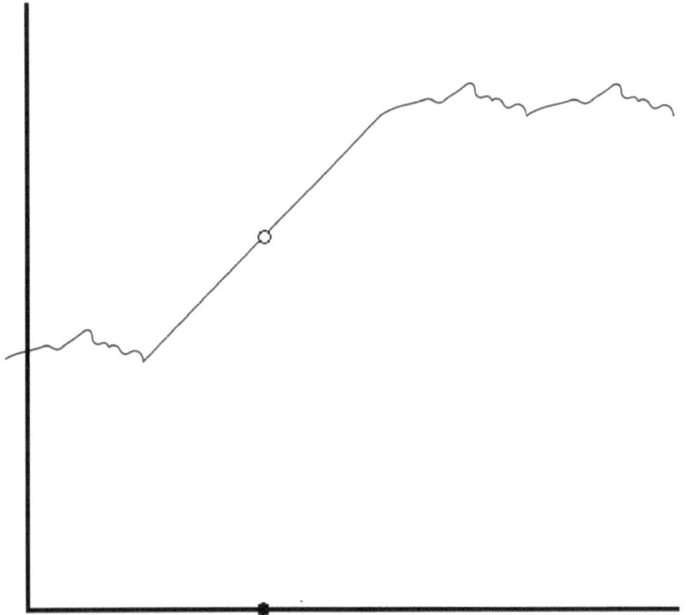

2. For every *x* where the mapping is defined there is only one corresponding value of *y*, so this is a function.
   Its graph looks like the following.

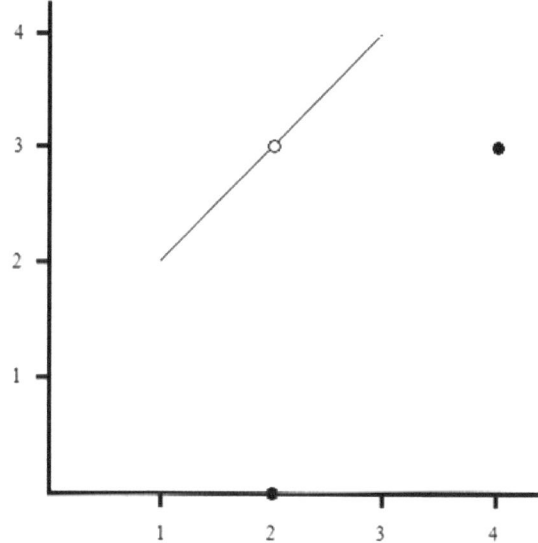

Page 51

1. Suppose $\lim\limits_{x \to 0} \frac{1}{x} = c$ where $c \geq 0$. Given ε, select $\delta < \frac{1}{c+\varepsilon}$. Given that c is the limit we must have that

   whenever $\qquad |x - 0| < \delta \qquad$ (or $x < \delta$ for positive x)

   then $\qquad\qquad |y - c| < \varepsilon$

Picking positive $x < \delta$, we have $x < \frac{1}{c+\varepsilon}$. Then $\frac{1}{x} > c + \varepsilon$.

Finally, $\quad |y - c| = \left|\frac{1}{x} - c\right| > |c + \varepsilon - c| = \varepsilon$, contradicting the requirement above for c to be the limit. Therefore, if the limit exists it can only be negative. A very slightly modified argument shows the limit cannot be negative. We conclude there is no limit.

Secondary problem for the reader: where in the above proof did we implicitly use the fact that $c \geq 0$? The answer is in the footnote.[6]

---

[6] If c were negative, then for sufficiently small ε we would find that δ would be negative.

2. The theorem cannot be applied when $b = 0$ because the first and second limits do not exist.

   The theorem cannot be applied when $b = 1$ because the first and third limits do not exist.

   The theorem can be applied when $b = 2$ because all three limits exist.

3. The theorem cannot be applied when $b = 0$ because the second and third limits do not exist.

   The parenthesized expression on the left evaluates to 0 for $x \neq 0$. We are taking the limit of the function h defined by

   $$h(x) = 0 \qquad \text{if } x \neq 0$$

   $$h(0) \qquad \text{is left undefined}$$

   This is a constant function that is undefined at one point. It does have a limit at $x = 0$ and the value of that limit is 0.

Page 59

1. We want to show f is continuous at the point 0. We know $f(0) = 0$ so the right side of the Equation of Continuity is 0. We now show that the limit on the left side has the value 0.

   For any ε we select $\delta = \varepsilon$. The task is to show that for all x such that $|x - a| < \delta$ it is also the case that $|f(x) - f(a)| < \varepsilon$. Plugging the specifics of this problem into these inequalities and using our choice of $\delta = \varepsilon$, we must show that for all x, whenever

$$|x| < \varepsilon$$

it is also the case that

$$|f(x)| < \varepsilon$$

When x is rational we have $f(x) = x$, so if the first inequality is satisfied so is the second.

When x is irrational we have $f(x) = 0$, so the second inequality is certainly satisfied.

This proves that

$$\lim_{x \to 0} f(x) = 0$$

Therefore

$$\lim_{x \to 0} f(x) = f(0)$$

which is the Equation of Continuity at $x = 0$, so f is continuous at this one point.

2. Pick a rational number $r = \dfrac{a}{b}$. Then $f(r) = \dfrac{1}{b}$. In order for f to be continuous at r the limit on the left side of the Equation of Continuity must exist and must have the right value. Since $f(r) = \dfrac{1}{b}$ the right value is $\dfrac{1}{b}$.

The difference between the value of f at irrational points and its value at r is the difference $\left|0 - \dfrac{1}{b}\right|$ which is just $\dfrac{1}{b}$. We want to select ε smaller than this, so let us take $\varepsilon = \dfrac{1}{2b}$.

If it were true that $\lim\limits_{x \to r} f(x)$ existed and that $\lim\limits_{x \to r} f(x) = \dfrac{1}{b}$, then for our chosen value $\varepsilon = \dfrac{1}{2b}$ we would be able to find some sufficiently small δ such that if

$$|x - r| < \delta$$

then

$$\left|y - \frac{1}{b}\right| < \frac{1}{2b}$$

But regardless of how small δ is selected, there are irrational numbers a closer to r than the distance δ, i.e., $|a - r| < \delta$. For these a we have $y = 0$ which means that

$$\left|y - \frac{1}{b}\right| = \left|0 - \frac{1}{b}\right| = \frac{1}{b} > \frac{1}{2b}$$

which contradicts the preceding inequality on y.

We conclude that the limit on the left side of the Equation of Continuity does not exist with the value $\dfrac{1}{b}$ (in fact it does not exist with any value when x is rational). The Equation of Continuity is not satisfied at r and the function is discontinuous at the rational points.

Page 66

1. $\dfrac{dy}{dx} = \lim\limits_{\Delta x \to 0} \dfrac{(x+\Delta x)^3 - x^3}{\Delta x}$

    $= \lim\limits_{\Delta x \to 0} \dfrac{3x^2\Delta x + 3x(\Delta x)^2 + (\Delta x)^3}{\Delta x}$

    $= \lim\limits_{\Delta x \to 0} 3x^2 + 3x\Delta x + (\Delta x)^2$

    $= 3x^2$

2. Using the definition of the derivative we have

    $$\dfrac{dy}{dx} = \lim\limits_{\Delta x \to 0} \dfrac{\sin(x+\Delta x) - \sin x}{\Delta x}$$

    Using the trigonometric identity

    $$\sin(a+b) = \sin a \, \cos b + \sin b \, \cos a$$

    we get

    $$\sin(x+\Delta x) = \sin x \, \cos \Delta x + \sin \Delta x \, \cos x$$

    Inserting this in the expression for the derivative yields

    $$\dfrac{dy}{dx} = \lim\limits_{\Delta x \to 0} \dfrac{\sin x \, \cos \Delta x + \cos x \, \sin \Delta x - \sin x}{\Delta x}$$

    $$= \lim\limits_{\Delta x \to 0} \dfrac{\sin x \, (\cos \Delta x - 1) + \cos x \, \sin \Delta x}{\Delta x}$$

$$= \sin x \left(\lim_{\Delta x \to 0} \frac{\cos \Delta x - 1}{\Delta x} \right) + \cos x \left(\lim_{\Delta x \to 0} \frac{\sin \Delta x}{\Delta x} \right)$$

where we used the limit theorems on page 49. In particular, because Δx is the variable in the limits and because $\sin x$ and $\cos x$ are constants with respect to Δx, we are allowed to take the factors $\sin x$ and $\cos x$ outside the limits.

3. Using the definition of the derivative we have

$$\frac{dy}{dx} = \lim_{\Delta x \to 0} \frac{\cos(x + \Delta x) - \cos x}{\Delta x}$$

Using the trigonometric identity

$$\cos(a + b) = \cos a \, \cos b - \sin a \, \sin b$$

we get

$$\cos(x + \Delta x) = \cos x \, \cos \Delta x - \sin x \, \sin \Delta x$$

Inserting this in the expression for the derivative yields

$$\frac{dy}{dx} = \lim_{\Delta x \to 0} \frac{\cos x \, \cos \Delta x - \sin x \, \sin \Delta x - \cos x}{\Delta x}$$

$$= \lim_{\Delta x \to 0} \frac{\cos x \left(\cos \Delta x - 1 \right) - \sin x \sin \Delta x}{\Delta x}$$

$$= \cos x \left(\lim_{\Delta x \to 0} \frac{\cos \Delta x - 1}{\Delta x} \right) - \sin x \left(\lim_{\Delta x \to 0} \frac{\sin \Delta x}{\Delta x} \right)$$

1. Use $x = 1$ as a base. Evaluating y and $\frac{dy}{dx}$ at $x = 1$ gives

$$y = 1 \quad \text{and} \quad \frac{dy}{dx} = 2x = 2 \times 1 = 2$$

Set $dx = 0.001$. Compute dy by

$$dy = \left(\frac{dy}{dx}\right) \times dx = 2 \times 0.001 = 0.002$$

The differential approximation to the function value at $x - 1.001$ is given by

$$y + dy = 1 + 0.002 = 1.002$$

The actual value is obtained by squaring 1.001:

$$1.001^2 = 1.002001$$

2. The exact value to four digits is $y = 2.09^3 = 9.1293$.

Use $x = 2$ as the base. Evaluating y and $\frac{dy}{dx}$ at $x = 2$ gives

$$y = 8 \quad \text{and} \quad \frac{dy}{dx} = 3x^2 = 3 \times 2^2 = 12$$

Set $dx = 0.09$. Compute dy by

$$dy = \left(\frac{dy}{dx}\right) \times dx = 12 \times 0.09 = 1.08$$

Then the differential approximation to the function value at $x = 2.09$ and its error are given by

$$y + dy = 8 + 1.08 = 9.08$$

$$e = 9.1293 - 9.08 = 0.0493$$

Now use $x = 2.1$ as the base. Evaluating y and $\frac{dy}{dx}$ at $x = 2.1$ gives

$$y = 9.261 \quad \text{and} \quad \frac{dy}{dx} = 3x^2 = 3 \times 2.1^2 = 13.23$$

Set $dx = -0.01$. Compute dy by

$$dy = \left(\frac{dy}{dx}\right) \times dx = 13.23 \times -0.01 = -0.1323$$

Then the differential approximation to the function value at $x = 2.09$ and its error are given by

$$y + dy = 9.261 - 0.1323 = 9.1287$$

$$e = 9.1293 - 9.1287 = 0.0006$$

The function is shown below in blue and the tangent lines providing the differential approximations are shown in red.

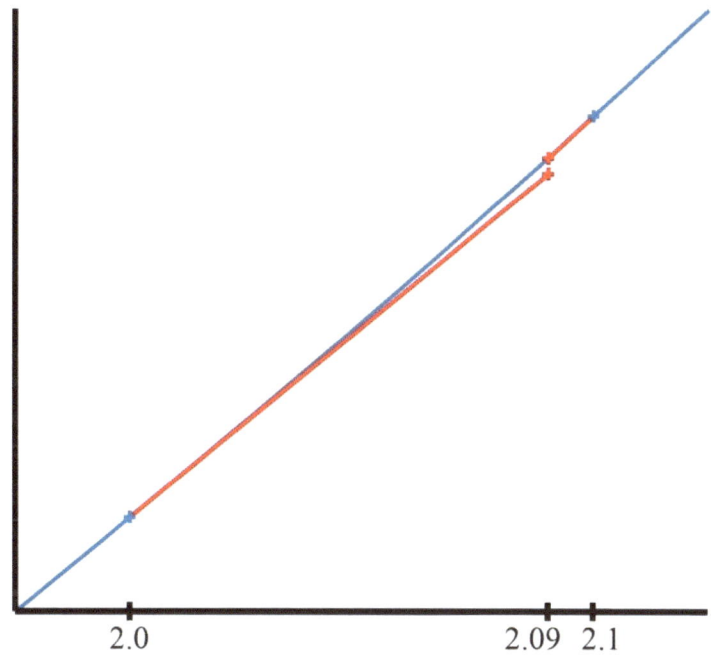

2.0   2.09 2.1

3. Convert $1°$ to radians:

$$1° \times \tfrac{\pi}{180°} = \tfrac{\pi}{180}$$

Use $x = 0$ as the base. Evaluating y and $\tfrac{dy}{dx}$ at $x = 0$ gives

$$y = \sin 0 = 0 \quad \text{and} \quad \tfrac{dy}{dx} = \cos 0 = 1$$

Set $dx = \tfrac{\pi}{180}$. Compute dy by

$$dy = \left(\tfrac{dy}{dx}\right) \times dx = 1 \times \tfrac{\pi}{180} = 0.017453$$

Then the differential approximation to the function value at $x = 1°$ is given by

$$y + dy = 0 + 0.017453 = 0.017453$$

Using a calculator gives us $\sin 1° = 0.017452$.

Table of Symbols

The numbers in parentheses are the pages where the symbols are first used. Note: Some of these symbols have additional meanings in more advanced texts.

$\frac{dy}{dx}$ (9) derivative of the function $y = f(x)$

$\pm a$ (10) denotes the pair of numbers $+a$ and $-a$

$| \; |$ (14) $|x|$ is the absolute value of x

$(\;)$ (18) (a,b) denotes the interval from a to b excluding the endpoints a and b. It can also mean the xy-coordinates of a point.

$[\;]$ (18) $[a,b]$ denotes the interval from a to b including the endpoints a and b

$\in$ (21) set inclusion; $a \in (p,q)$ means that a is in the interval (p,q)

$x \rightarrow b$ (27) variable x approaches (arbitrarily closely) the number b

$\sqrt[3]{a}$ (29) cube root of a

$\lim_{x \rightarrow b} y$ (29) the limit of y as x approaches b

δ (31) maximum deviation of variable x from its limiting value

ε (31) maximum deviation of variable y from its limiting value

Δy (61) increment of a variable, here variable y

dy (67) differential of a variable, here variable y

$\approx$ (70) used in place of an equals sign (=), this means that the values on its left and right sides are approximately equal

Glossary

The numbers in parentheses are the pages where the words are first used. Note: Some of the terms defined here have different meanings in more advanced mathematics, e.g., both "algebra" and "linear function".

Absolute value

(14) The distance of a number from zero, expressed as a non-negative number. The absolute value of -3 is $+3$.

aka

(44) This stands for "also known as". It asserts that what follows it is another name for what came before it. For example, "We want the value of y aka $f(x)$."

Algebra

(1) The branch of mathematics that uses symbols to represent unknown numbers, solves equations to obtain the values of those symbols, and studies the relations between the symbols.

Analytic geometry

(1) The study of geometry in 2 or 3 dimensions using a coordinate system.

Calculus

(3) The branch of mathematics dealing with calculating rates of change and areas.

Cauchy, Augustin-Louis

(4) 1789-1857 French mathematician who pioneered rigor and completely redesigned the foundations of calculus. By modern standards his creation contained a few gaps that were later filled in by Weierstrass.

Constant function

(52) Any function that always takes on the same value regardless of the value of x.

Continuity

(4) A mathematical concept with a precise definition that captures the intuitive concept that there are no breaks in the graph of a function.

cos

(65) A trigonometric function of angle defined as a certain ratio of sides (a different ratio than used

	by the sin function) of a right triangle having that angle.
Cube	(29) The number that is the result of multiplying three instances of another number. For example, the cube of 2 is 8 (2x2x2) and the cube of 3 is 27 (3x3x3).
Cube root	(29) Given a number A, the cube root is the number whose cube is A. For example, the cube root of 8 is 2, the cube root of 27 is 3 and the cube root of 13 is an infinite decimal that starts 2.351…
d'Alembert, Jean	(4) 1717-1783 French mathematician who was one of the early developers of calculus.
Delta	(31) The fourth letter of the Greek alphabet, written δ in lower case and Δ in upper case.
Derivative	(3) A derived function whose value is the slope of an original function.
Differential	(67) An increment in y or x, mutually defined in such a way that their ratio equals the derivative.
e.g.	(18) Abbreviation of the Latin *exempli gratia*, meaning "for the sake of example" or, more colloquially, "for example". It indicates that what follows shows an example of what came before but is not a full restatement of what came before.
Epsilon	(31) The fifth letter of the Greek alphabet, written ε in lower case.
Equation of Continuity	(5) The equation that provides the rigorous mathematical definition of continuity.
Explicit	(54) Being plainly stated so as to leave no doubt.
Formula	(10) A mathematical expression that has a numeric value when numbers are assigned to all its variables, e.g., $x^2 - 2$.

Function	(10) A mapping (see definition) restricted so that each number in the first set has only one corresponding number in the second set.
Holistic	(67) Dealing with something in its entirety while ignoring any structure it may have.
i.e.	(14) Abbreviation of the Latin *id est* meaning "that is". This is the same meaning as the English alternative "in other words". It indicates that what follows is a full restatement of what came before.
Identity function	(53) The function $y = x$ or, equivalently, $f(x) = x$.
Implicit	(53) Understood without being stated.
Infinitesimal	(8) A type of number that is not defined in standard mathematics. Infinitesimals are smaller than all positive numbers yet are larger than zero.
Interval	(18) The set of all numbers between two given numbers which are called the endpoints. An interval may or may not include its endpoints.
Irrational number	(55) A number that cannot be expressed as the ratio of two integers. Rational and irrational numbers are intimately mixed in the number line.
Leibniz, Gottfried	(4) 1646-1716 German polymath (an expert in many fields) who independently invented calculus.
Limit	(4) The unique number a function can be made to approach arbitrarily close to as the variable it depends on approaches some predetermined number.
Linear function	(17) A function of the form $mx + b$, where m and b are constants. They are called "linear" because their graphs are straight lines.

Mapping	(10) A correspondence from each element of one set to one or more elements of a second set.
Newton, Isaac	(4) 1642-1727 English mathematician and physicist who independently invented calculus.
Pedagogical	(ix) Related to teaching.
Polynomial	(54) A function that involves only sums of powers of a variable, each power multiplied by some constant.
Rational number	(55) A number that can be expressed as the ratio of two integers, although it may also be expressed in other ways. Rational and irrational numbers are intimately mixed in the number line.
Robinson, Abraham	(9) 1918-1974 German-American mathematician who created the non-standard formulation of calculus.
Root	(17) A value of x that makes $f(x)$ evaluate to zero. For example, the roots of $f(x) = x^2 - 4x + 3$ are 1 and 3.
Secant	(61) A straight line that intersects a curve in at least two points.
sin	(65) A trigonometric function of angle defined as a certain ratio of sides (a different ratio than used by cos) of a right triangle having that angle.
Slope	(7) The rate of change of one variable relative to another.
Tangent	(7) A straight line that intersects a curve in just one point and which does not cross the curve at that point.
Theorem	(49) Usually a mathematical statement that has been proven but sometimes a mathematical

statement that has been conjectured but not proven.

Trigonometry

(65) The branch of mathematics that deals with the lengths of a triangle's sides and with its angles.

Weierstrass, Karl

(4) 1815-1897 German mathematician who created the completely rigorous, modern formulation of calculus.

Index

www.ingramcontent.com/pod-product-compliance
Lightning Source LLC
Chambersburg PA
CBHW050851180526
45159CB00007B/2641

* 9 7 8 0 9 9 9 9 6 1 7 4 0 3 *